Drug Testing in Alternate Biological Specimens

F O R E N S I C
SCIENCE AND MEDICINE

DRUG TESTING IN ALTERNATE BIOLOGICAL SPECIMENS

Edited by

Amanda J. Jenkins

Lake County Crime Laboratory, Painesville, OH

Foreword by

Yale H. Caplan

National Scientific Services, Baltimore, MD

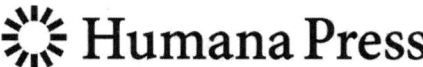 Humana Press

Editor
Amanda J. Jenkins
Lake County Crime Laboratory
Painesville, OH

ISBN: 978-1-58829-709-9   e-ISBN: 978-1-59745-318-9

Library of Congress Control Number: 2007940757

Printed on acid-free paper

9 8 7 6 5 4 3 2 1

springer.com

This work is dedicated to the global community of toxicologists, analytical chemists, and other scientists who have contributed to the body of knowledge in the field of forensic toxicology.

Amanda J. Jenkins, 2007

We must not forget that when radium was discovered no one knew that it would prove useful in hospitals. The work was one of pure science. And this is a proof that scientific work must not be considered from the point of view of the direct usefulness of it. It must be done for itself, for the beauty of science, and then there is always the chance that a scientific discovery may become like the radium a benefit for humanity.

Marie Curie (1867–1934)

Lecture at Vassar College, Poughkeepsie, NY, USA

May 14, 1921

Foreword

Forensic toxicology encompasses the analysis for drugs and chemicals including the most common drugs of abuse and also focuses on the interpretation, that is, the understanding and appreciation of the results of this testing in a medical–legal context. The same methods and principles can also be applied to clinical situations. Traditionally, forensic toxicology focuses on postmortem investigation, workplace drug use assessment, and human performance evaluation, but in many instances, clinical testing becomes forensic when treatment is associated with a court order or family situations lead to custody struggles. The results of toxicology testing are often presented to courts for the adjudication of an issue but are very often misunderstood or worse misrepresented. We need to remember that a test is not a test. A test result is only as good as the question it is asked to answer. Toxicology test results must, therefore, be introduced by qualified toxicologists.

The traditional specimens used in testing include blood or its component parts, that is, plasma or serum, and urine. This is in part because these are the easiest to collect. In addition, in the case of blood, or its components, it represents the dynamic state of drug distribution in the body with the best relation to the state of the individual's pharmacologic condition (therapeutic, impairment, and death). In the case of urine, we have a static fluid that generally does not correlate with the pharmacological effects in an individual, rather it represents high concentrations of drugs and metabolites and demonstrates prior use. Thus, the ready accessibility and knowledge of the pharmacokinetics and distribution of drugs caused toxicologists to focus on these specimens. Further they were within the limits of the known analytical testing methodology.

Although drug testing includes many hundreds of prescription drugs, illicit drugs, or other chemicals, five classes of drugs are common to all forensic arenas. These are the amphetamines (including amphetamine and methamphetamine), cocaine, marijuana, narcotics (including morphine, codeine, and others), and phencyclidine.

Testing methodology has continually evolved now including GCMS, GCMSMS, LCMS, and LCMSMS improving sensitivity and reducing sample sizes, thus permitting effective analysis of additional specimens that were previously inaccessible. These non-traditional materials may be summarized into three groups:

1. Clinical ante mortem specimens including amniotic fluid, breast milk, and meconium.

2. Postmortem specimens to facilitate death investigations including vitreous humor, brain tissue, liver tissue, bones, bone marrow, hair and nails.
3. Workplace testing enhancement including oral fluid (saliva), hair, and sweat.

The chapters in this book focus on these less traditional specimens and particularly the application of these areas of practice to the drugs of abuse. The use of these specimens enhances the forensic investigation and leads to a more complete understanding of the drug-related event. The sum purpose of all toxicological testing is to insure the determination of the cause of drug deaths, the impairment of individuals by drugs, and/or an individual's prior use of drugs. All specimens have a specific formation and time line. The incorporation of drugs into or out of a specimen is a function of the drugs chemical structure, pharmacokinetics, and the nature of the time line for the specimen. Specimens have similarities and differences, hence, strengths and limitations. Each provides a unique historical picture. Results between all specimens do not have to agree (i.e., they all need not be positive at the same time). Understanding the differences is essential to interpretation and one of the purposes of this book.

The term alternate matrices connotes that specimens in addition to the traditional matrices may be useful in diagnosis, particularly if and when the traditional matrices are not available or are contaminated. However, more frequently the specimens should be considered "complimentary," that is, they can confirm, enhance, or facilitate interpretation of the results from the traditional matrices. As for all drugs and specimens, the process of interpretation should include consideration of all aspects of the investigation, including the analysis of multiple specimens.

For example:

- Testing vitreous humor particularly in alcohol cases may overcome the issue of postmortem redistribution.
- Testing brain, liver, and hair or nails may be useful in decomposed bodies where blood and urine are not available.
- Testing oral fluid and hair in the workplace may contribute to evaluating the frequency of use and/or to overcome adulteration of urine.
- Testing maternal specimens and meconium may allow assessment of substance abuse against newborns where sufficient volumes of traditional specimens are unavailable.

Some highlights of the book include:

- The liver is the largest organ in the human body and is relatively unaffected by postmortem redistribution as compared with blood.
- Brain is useful in the interpretation of time intervals between administration of drug and death.
- The composition of amniotic fluid and breast milk and the mechanisms known to effect drugs of abuse transfer to these matrices are reviewed.
- Saliva or oral fluid is discussed with regard to the effect of route of administration, collection procedures, and saliva : plasma ratios on the amount of drug deposited.

- Sweat as a biological matrix is described including an overview of the structure of the skin, the composition and production of sweat, and the approaches used to collect sweat.
- Bone and bone marrow are facilitated as specimens following extraction by soaking bone in organic solvent and subjecting to routine drug assays.
- Meconium may provide a history of in utero drug exposure. Although easy to collect, small sample sizes, lack of homogeneity, different metabolic profiles, and the requirement for low-level detection present analytical challenges.
- The utility of nails is examined reviewing the basic structure of the nail, mechanisms of drug incorporation, analytical methodologies, and interpretation of results.
- Vitreous humor is reviewed considering pertinent studies that have examined drug deposition into the specimen. Discussion includes the increased stability of certain drugs in this matrix and its amenability to analysis with little or no pretreatment.
- The chapters offer windows into the wider world of drug testing. They provide the chance to go further to unfold new forensic mysteries and answer new questions for the criminal justice system.

Yale H. Caplan, Ph.D., D-ABFT
National Scientific Services, Baltimore, MD, USA

Preface

Drug abuse in the developed world is an international problem. In the USA, in an effort to deter drug use and identify abusers so they may receive treatment, testing an individual's urine has become a large commercial enterprise. Drug testing has also been a traditional part of clinical care in medicine and in the medicolegal investigation of death. While scientists conducting drug testing in the postmortem arena routinely analyze a variety of biological matrices, the specimen of choice in the drug testing industry in the USA is urine and in clinical medicine, serum. In recent years, interest has grown in the use of other matrices as drug testing media. Although many peer-reviewed articles have appeared in the scientific literature describing drug appearance in these "alternate" biological specimens, the field is without a general text summarizing the state of our knowledge.

The objective of this book is to provide forensic toxicologists with a single resource for current information regarding use of alternate matrices in drug testing. Where appropriate information provided includes an outline of the composition of each matrix, sample preparation and analytical procedures, drugs detected to date, and a discussion of the interpretation of positive findings. As many compounds could potentially be discussed, the focus of this work is drugs of abuse to include amphetamines, cannabinoids, cocaine, opioids, and phencyclidine. Each chapter is written by an authors(s) with familiarity in the subject, typically, by conducting research and casework using the specimen discussed and publishing in peer-reviewed journals.

Amanda J. Jenkins, **Ph.D., D-ABC, D-FTCB**

Contents

Contributors

GAIL COOPER • Forensic Medicine and Science, University of Glasgow, Glasgow, UK

OLAF H. DRUMMER • Victorian Institute of Forensic Medicine and Department of Forensic Medicine, Monash University, Southbank, Australia

NEIL A. FORTNER • ChoicePoint, Inc., Keller, TX, USA

DIANA GARSIDE • Chapel Hill, NC, USA

BRUCE A. GOLDBERGER • Department of Pathology, Immunology & Laboratory Medicine, Department of Psychiatry, Gainesville, University of Florida College of Medicine, FL, USA

GRAHAM R. JONES • Office of the Chief Medical Examiner, Edmonton, Alberta, Canada

REBECCA A. JUFER • Baltimore, MD, USA

SARAH KERRIGAN • College of Criminal Justice and Department of Chemistry, Sam Houston State University, Huntsville, TX, USA

PASCAL KINTZ • Laboratoire ChemTox, Illkirch, France

BARRY S. LEVINE • Baltimore, MD, USA

CHRISTINE M. MOORE • Toxicology Research and Development, Immunalysis Corporation, Pomona, CA, USA

PETER P. SINGER • Office of the Chief Medical Examiner, Edmonton, Alberta, Canada

VINA SPIEHLER • Spiehler and Associates, Newport Beach, CA, USA

THOMAS STIMPFL • Institute of Legal Medicine, University of Hamburg, B-Eppendorf, Hamburg, Germany

Chapter 1

Specimens of Maternal Origin: Amniotic Fluid and Breast Milk

Sarah Kerrigan and Bruce A. Goldberger

Summary

This chapter describes the composition of amniotic fluid and breast milk and the mechanisms known to effect drug transfer to these matrices. Drugs-of-abuse detected in these specimens and discussed in this chapter include cocaine and metabolites, phencyclidine (PCP), benzodiazepines, barbiturates, opioids, amphetamines and cannabinoids.

Key Words: Amniotic fluid, breast milk, drugs-of-abuse.

1. INTRODUCTION

The physical and chemical characteristics of a drug and biofluid can be useful predictors of drug transfer into various compartments of the body. The principal mechanism of transfer for most drugs is passive diffusion. The pKa, lipid solubility, protein binding, and body fluid composition largely determine the extent to which the drug is present. Physico-chemical characteristics of selected drugs are given in Table 1. Transfer of drugs from the circulating blood (pH 7.4) to another biological fluid involves transport across membranes that are an effective barrier against ionized, highly polar compounds. Following penetration of the membrane and transfer into the biofluid, the pH differential may result in ionization of the drug, restricting further mobility. Accumulation of the drug in this way is commonly referred to as "ion trapping."

From: *Forensic Science and Medicine: Drug Testing in Alternate Biological Specimens*
Edited by: A. J. Jenkins © Humana Press, Totowa, NJ

1

Table 1
Properties of Selected Drugs

Drug	pKa	V_d (L/kg)	Log P	Fb (%)	$T_{1/2}$
Acetaminophen	9.5	1	0.5	25	1–3 h
Alprazolam	2.4	0.7	2.12	70–80	11–15 h
Amitriptyline	9.4	15	4.94	91–97	9–36 h
Caffeine	14.0, 10.4	0.5	−0.07	35	2–10 h
Cocaine	8.6	1–3	2.3	92	0.7–1.5 h
Diazepam	3.3	0.5–2.5	2.7[a]	98–99	20–40 h
Diphenhydramine	9.0	4.5–8	3.3	80	2.4–9.3 h
Fentanyl	8.4	4	2.3[a]	80	3.7 h
Fluoxetine	9.5	27	4.05	95	4–6 days
Ketamine	7.5	4	3.1	20–50	2–3 h
Meperidine	8.7	4	2.7	50–60	3–6 h
Methadone	8.3	4	3.93	92	10–25 h
Methamphetamine	10.1	3–7	2.1	10–20	9 h
Morphine	8.1	3–5	–	20–35	2–3 h
Oxycodone	8.9	–	0.7	87–94	2–3 h
Phencyclidine (PCP)	8.5	6	4.7	65–80	7–46 h
Phenobarbital	7.4	0.5	1.5	50	90–100 h
Phenytoin	8.3	0.5–1.2	2.5	90	7–42 h
Propoxyphene	6.3	16	4.2	70–80	8–24 h
Salicylic Acid	3.0, 13.4	0.1–0.2	2.3	40–80	2–4 h
Sertraline	9.5	20	5.29	98	26 h
Temazepam	1.6	1	2.19	96	8–15 h
Δ-9-tetrahydrocannabinol	10.6	10	7.6	94–9%	2 h
Valproic Acid	4.6	0.1–0.2	2.8	90	6–20 h

Fb, fraction bound to plasma protein; Log P, partition coefficient (octanol/water); pKa, dissociation constant; $T_{1/2}$, plasma half life; V_d, volume of distribution.
[a]Partition coefficient in octanol/pH 7.4 buffer.

Source: *Clarke's Analysis of Drugs and Poisons*, 3rd Edn, AC Moffatt, MD Osselton and B Widdop, Eds. Pharmaceutical Press, London, UK, 2004. *Disposition of Toxic Drugs and Chemicals in Man*, 7th Edn, RC Baselt, Biomedical Publications, Foster City, CA, 2004.

Table 2 summarizes the effect of pH and pKa on acidic, basic, and neutral drugs, and Fig. 1 illustrates the concept of ion trapping.

The increasing use of illegal drugs by expectant mothers has led to an increased need for prenatal toxicological testing. Exposure to drugs-of-abuse may result in higher rates of congenital anomalies and neonatal complications. Identification of gestational drug exposure may benefit the newborn in terms of increased vigilance and monitoring of the infant by medical and social services. However, amniotic fluid and breast milk are not routinely used to determine maternal drug use. Other samples of maternal origin such as urine, saliva,

Table 2
Effect of pH and pKa on Acidic, Basic, and Neutral Drugs

	pH units from pKa				
	−2	−1	0	+1	+2
Drug type	% Ionized drug				
Acidic drugs: acetaminophen, ampicillin, barbiturates, NSAIDs, phenytoin, probenecid, and THC metabolites	1	9	50	91	99
Neutral drugs: carbamazepine, glutethimide, meprobamate	0	0	0	0	0
Basic drugs: amphetmines, antiarrhythmias, antidepressants, antihistamines, cocaine, narcotic analgesics, PCP, and phenothiazines	99	91	50	9	1

NSAIDs, non-steroidal anti-inflammatory drugs; THC, tetrahydrocannabinol; PCP, phencyclidine.

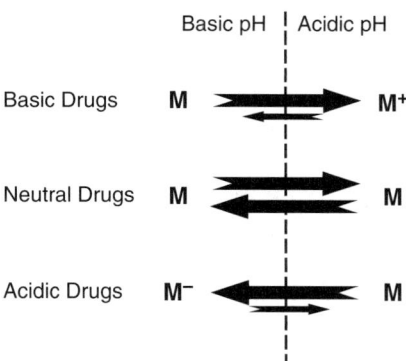

Fig. 1. Dissociation of Drugs and Ion Trapping.

blood, or hair can be used for this purpose during pregnancy. More importantly, the detection of drugs or drug metabolites in amniotic fluid and breast milk are essential to our understanding of the pharmacokinetic principles governing intrauterine and prenatal mechanisms of drug transfer. Factors that influence specimen selection is listed in Table 3, and the advantages and disadvantages of each are summarized in Table 4.

1.1. Rates of Drug Use

According to self-reported data, it is estimated that almost 4% of pregnant women aged 15–44 years have used illicit drugs *(1)*. Marijuana was the

Table 3
Factors Influencing Biological Specimen Selection

Sample Collection

Invasiveness
Risk of infection, complication, hazards
Protection of privacy
Ease and speed of collection
Training of personnel (medical/non-medical)
Likelihood of adulteration
Contamination
Volume of specimen

Analysis

Qualitative or quantitative results
Window of detection
Drug concentration/accumulation in biofluid
Parent drug or metabolite(s)
Stability of drug(s)
Biofluid storage requirements
Pretreatment of specimen
Limitations of the matrix
Likelihood of interferences
Inter and intrasubject variability of the matrix
Use of existing analytical procedures
Speed of analysis
Personnel training requirements
Appropriate cut-off concentrations

Interpretation

Pharmacologic effects
Indicator of recent drug use (hours)
Short-term drug exposure (days)
Long-term drug exposure (weeks)
Forensic defensibility

most widely used drug (2.8%) followed by non-medical use of prescription drugs (0.9%). Other drugs used included cocaine, heroin, inhalants, and hallucinogens including phencyclidine (PCP) and lysergic acid diethylamide (LSD). Furthermore, the percentage reporting past month use of an illicit drug was only marginally lower for pregnant women aged 15–17 (12.9%) compared with non-pregnant women in the same age group (13.5%) *(2)*.

Table 4
Advantages and Disadvantages of Amniotic Fluid and Breast Milk

Biofluid	Advantages	Disadvantages
Amniotic fluid	Minimal sample preparation	Highly invasive sampling procedure
	Amenable to most analytical techniques	Requires local anesthetic, ultrasound scan and highly trained medical personnel
	Relatively few interferences	Risk of complication associated with sampling
	Useful in determining intrauterine drug exposure at an early stage of development	
Breast Milk	Many drugs determined	High lipid content may interfere with analysis
	Maternal and neonatal drug exposure can be determined	Additional extraction steps may be required
		Disposition of drug varies with milk composition
		Matrix variability between individuals and in one feed
		Inconvenient specimen collection (requires pump)
		Invasion of privacy

The accuracy of self-reported drug use is questionable because of the stigma of drug use in pregnancy and the associated legal and ethical issues. Toxicological testing of maternal or neonatal specimens for commonly abused drugs may provide a more realistic estimate of drug use. In these studies, drug prevalence can vary dramatically, depending on whether the population is considered high or low risk *(3)*. The prevalence of drug abuse among pregnant women throughout the USA is reported to be between 0.4 and 27% *(4)*. In one study of newborn drug screening among a high-risk urban population, rates of cocaine, morphine, and cannabinoid use were 31, 21, and 12% respectively *(5)*. However, this study likely overestimates drug use because results were determined by radioimmunoassay and were not confirmed by another technique. More recently, the US National Institute on Child Health and Development undertook a multi-site study of more than 8000 infants. Newborn drug testing of meconium samples revealed that 10.7% of the samples contained cocaine, opiates, or both *(6)*. In this study, positive immunoassay test results were confirmed by a secondary technique.

1.2. Drug Effects

The use of drugs during pregnancy may pose a potential risk to both the mother and the fetus. Drug effects on fetal safety are generally evaluated using animal data or available experience from human pregnancies. Based on this approach, the United States Food and Drug Administration established a categorization of drugs according to safety in 1979, which is currently under revision *(7)*. Although such categorizations provide a rough estimate for adverse fetal consequences, they are often derived from very limited data sets *(8)*. In a recent study of prescription drug use in pregnancy, an estimated 64% of women were prescribed a drug other than a vitamin or mineral supplement prior to delivery *(9)*, and as many as 40% received a drug during delivery from category C (drugs for which human safety during pregnancy has not yet been established) according to the FDA classification system.

Pre-natal and intrauterine drug exposure remains a major health concern. Numerous effects including intrauterine growth retardation, birth defects, altered neurobehavior, and withdrawal syndromes have been reviewed *(3,10, 11)*. Pre-natal cocaine use is associated with placental abruption and premature labor, whereas intrauterine cocaine exposure is associated with prematurity, microcephaly, congenital anomalies, necrotizing enterocolitis, and stroke or hemorrhage. Amphetamines may lead to complications similar to those of cocaine-exposed infants, including increased rates of maternal abruption, prematurity and low birth weight. Heroin use during pregnancy has been associated with low birth weight, miscarriage, prematurity, microcephaly, and intrauterine growth retardation. Marijuana has been associated with visual anomalies and a number of persistent neurocognitive effects in the latter stages of development. Neonatal abstinence syndrome is associated with in utero exposure to opioids, cocaine, and methamphetamine. Drug effects on the developing organism are dependent on a number of factors including the activity and retention of the drug and its metabolites in the maternal–fetal compartment, as well as the dose and duration of exposure.

2. AMNIOTIC FLUID

2.1. Anatomy and Physiology

Amniotic fluid, produced by cells that line the innermost membrane of the amniotic sac (amnion), is the liquid that surrounds and protects the embryo during pregnancy. This fluid cushions the fetus against pressure from internal organs and from the movements of the mother. Production of fluid commences on the first week after conception and increases steadily until the 10th week, after which the volume of fluid rapidly increases. Amniotic fluid, which may total about 1.5 L at 9 months, contains cells and fat that may give the liquid a slightly cloudy appearance. The protein concentration and the pH of the fluid vary with gestational age. Amniotic fluid is constantly circulated, being swallowed by the fetus, processed, absorbed, and excreted by the fetal kidneys as urine at rates as high as 50 mL per hour. This circulation of fluid continuously exposes the fetus to compounds that may be absorbed in the gut or diffused through fetal skin in the early stage of development. The encapsulation of the fetus in this fluid may prolong exposure to harmful drugs and metabolites. The pharmacokinetics of drug disposition in utero varies from drug to drug, and the acute and chronic effects that may result are the topic of continuing research.

2.2. Drug Transfer

A number of maternal, fetal, and placental factors have been documented to affect fetal drug exposure. Of these, binding to serum proteins in the maternal and fetal circulation and fetal elimination are particularly important. Conjugated drug metabolites, which tend to be highly water soluble, may accumulate in the fetus or amniotic fluid because of limited placental transfer. The principle routes of drug transfer into amniotic fluid occur through the placenta and the excretion of water-soluble drugs into the fetal urine. The placenta is an extra embryonic tissue that is the primary link between the mother and the fetus. Passive diffusion and to a lesser extent active transport and pinocytosis are responsible for drug transfer. The physico-chemical properties of the drug, such as pKa, lipid solubility, and protein binding, largely influence the drug's ability to cross the placenta and enter the fetal circulation and the amniotic fluid. The amnion is considered to be a deep compartment, whereby equilibration with adjacent compartments is achieved relatively slowly.

Transplacental passage of small lipophilic drugs occurs readily and is limited only by blood flow rates. By comparison, the rate of transfer of a hydrophilic drug may be approximately one-fifth that of a lipophilic drug of similar size. With drugs that tend to be highly protein bound, only the small fraction of free drug may diffuse across the membrane. Small lipid-soluble drugs can rapidly diffuse across the placental barrier, producing similar drug concentrations in both amniotic fluid and fetal plasma. Larger, water-soluble compounds that are transferred more slowly are incorporated into the amniotic fluid through fetal urine. Basic drugs may accumulate in the amnion due to ion trapping, resulting in drug concentrations in excess of those found in fetal or maternal plasma.

Large, lipid-soluble drugs are more readily transferred to the fetus but less readily transferred to the amniotic fluid due to reabsorption in the fetal kidney. The fetal kidney is not an effective route of drug elimination because fetal renal blood flow is only 3% of the cardiac output compared with 25% in an adult. In addition to transplacental passage of metabolites from the mother, biotransformation by the immature fetal liver may also be responsible for the appearance of some drug metabolites in the amniotic fluid. The glucuronidation capacity of the adult liver is estimated to be 6- to 10-fold greater than fetal liver at 15–27 weeks *(12)*.

2.3. Sample Collection and Drug Analysis

The collection of amniotic fluid (amniocentesis) usually takes place between the 16th and 20th week of pregnancy. The liquid is usually collected to test for fetal abnormalities or to learn the sex of the child. The presence of illicit drugs or their metabolites in amniotic fluid suggests that the fetus has been exposed to these substances through maternal blood circulation. A maternal serum sample taken at the same time as the test may provide complementary toxicological data and help assess the relative risk to the fetus. Amniocentesis is an invasive procedure. Prior to the test, an ultrasound scan is used to determine the position of the fetus. A needle is inserted through the abdomen into the uterus where there is the least chance of touching the placenta or the fetus. Although complications are rare, miscarriage occurs in approximately 1% of women. Typically, 5–30 mL of amniotic fluid is removed. The pH of amniotic fluid decreases from slightly alkaline to near neutral pH at full term due to fetal urination.

Amniotic fluid, which is 99% water, contains dilute plasma components, cells, and lipids. Following amniocentesis, the fluid may be centrifuged and the supernatant layer frozen prior to drug testing. Drugs present in amniotic fluid can be analyzed using well-established techniques that are routinely used for blood, urine, or serum (Table 5). Relatively few interferences are encountered with amniotic fluid due to its high water content. Sample pre-treatment prior to an immunoassay screening test may not be necessary; however, confirmation by gas chromatography/mass spectrometry (GC/MS) requires isolation of the drug using liquid/liquid or solid phase extraction techniques. Although amniotic fluid is extremely amenable to routine toxicology tests, it is not routinely used for this purpose because of the invasiveness of the specimen collection procedure. Subsequently, there are considerably fewer toxicological studies compared with other specimens.

Table 5
Methods of Analysis of Drugs in Amniotic Fluid and Breast Milk

Purification/drug extraction

  Liquid–liquid extraction
  Solid-phase extraction
  Supercritical fluid extraction

Drug detection by immunochemical techniques

  Cloned enzyme donor immunoassay (CEDIA®)
  Enzyme-linked immunosorbent assay (ELISA)
  Enzyme-multiplied immunoassay technique (EMIT™)
  Fluorescence polarization immunoassay (FPIA®)
  Kinetic interaction of microparticles in solution (KIMS®)
  Radioimmunoasay (RIA)

Drug identification by chromatographic techniques

  Capillary electrophoresis (CE)
  Gas chromatography (GC)
  Gas chromatography/mass spectrometry (GC/MS)
  Gas chromatography/mass spectrometry/mass spectrometry (GC/MS/MS)
  High-performance liquid chromatography (HPLC)
  Liquid chromatography/mass spectrometry (LC/MS)
  Liquid chromatography/mass spectrometry/mass spectrometry (LC/MS/MS)
  Thin layer chromatography (TLC)

2.4. Toxicological Findings

Cocaine and benzoylecgonine concentrations in amniotic fluid from known cocaine users ranged from 0.4–5 to 0–0.25 mg/L *(13)*. Following the death of a pregnant woman, amniotic fluid cocaine and benzoylecgonine concentrations of 3.3 and 1.6 mg/L were reported *(14)*. After crossing the placental barrier by simple diffusion, the drug distributes between fetal and maternal blood. The amniotic sac and its contents serve as a deep compartment with restricted, slow equilibrium between adjacent compartments. As a result, amniotic fluid inside this protective sac may expose the fetus to potentially harmful drugs or metabolites that are sequestered in the biofluid. Animal studies have shown a three- to fourfold increase in cocaine concentration compared to fetal or maternal plasma. The concentration of benzoylecgonine in amniotic fluid was also shown to be higher than newborn urine *(13)*. Limited fetal hepatic function, ion trapping, and the slow pharmacokinetic exchange present in this deep compartment can increase intrauterine drug exposure and thus compound the risk of complications.

Other basic drugs, such as PCP and fentanyl, have also been detected. PCP was detected in amniotic fluid at a concentration of 3.4 ng/mL some 36 days after hospitalization of a pregnant woman (15). Animal studies have shown that PCP readily crosses the placenta and may concentrate in the fetal tissues. Fentanyl was detected in a series of 14 paired maternal serum and amniotic fluid samples taken during the first trimester (16). The drug was detected in amniotic fluid within 5 min of intravenous drug administration at concentrations that sometimes exceeded those found in maternal serum.

Narcotic analgesics are reported to cross the placental barrier rapidly. Following parenteral administration of morphine to the mother, the fetal–maternal ratio of morphine concentration in blood reached unity at 5 min (17). However, at physiological pH, narcotic analgesics tend to be predominantly charged, so in the absence of other factors, the concentration of the drug in the amniotic fluid is expected to be lower than that of the maternal plasma. In one case study however, fetal–maternal blood concentration ratios for morphine, 6-acetylmorphine (6-AM), and codeine were 4.86, 38, and 3.5, respectively (18). Concentrations of morphine, morphine-3-glucuronide, and 6-AM in amniotic

Table 6
Drug and Drug Metabolites Detected in Specimens of Maternal Origin

Biofliud	Drug/drug metabolite
Amniotic fluid	Benzoylecgonine, caffeine, cocaine, cocaethylene, diazepam, digoxin, ecgonine methyl ester, 2-ethylidine-3,3-diphenylpyrrolidine (EDDP), fentanyl, 6-acetylmorphine (6-AM), meperidine, methadone, morphine, morphine-3-glucuronide, nitrazepam, norcocaine, nordiazepam, phencyclidine (PCP), phenobarbital, secobarbital, and valproate (valproic acid)
Breast milk	Acetaminophen, amitriptyline, amphetamine, benzoylecgonine, chloral hydrate, citalopram, cocaine, cocaethylene, codeine, chloral hydrate, desipramine, desmethylcitalopram, diazepam, didesmethylcitalopram, dothiepin, doxepin, fentanyl, flunitrazepam, fluoxetine, hydromorphone, 11-hydroxy THC, imipramine, lorazepam, meperidine, morphine, methadone, nitrazepam, norcocaine, nordiazepam, nordoxepin, norfluoxetine, normeperidine, norsertraline, nortriptyline, oxazepam, oxycodone, paroxetine, salicylate, sertraline, temazepam, Δ9-tetrahydrocannabinol (THC), 11-nor-9-carboxy-Δ9-tetrahydrocannabinol (THCA), and trichloroethanol

fluid were 604, 209, and 128 µg/kg compared with 280, 801 and 4 µg/kg in the maternal blood of a 17-year old pregnant female who died of a heroin overdose. Mean methadone concentrations in maternal serum and amniotic fluid were 0.19 and 0.20 mg/L, respectively, following maternal drug use *(19)*. Methadone and 2-ethylidine-3,3-diphenylpyrrolidine (EDDP) concentrations of 0.66 and 0.52 mg/L in amniotic fluid were measured full-term in a female who was maintained on 110 mg methadone a day *(20)*.

Benzodiazepines cross the placenta because of their lipid solubility and lack of ionization. However, drug concentrations in the amniotic fluid remain relatively low due to extensive protein binding in the maternal plasma, minimal renal excretion by the fetus, and the absence of ion trapping. Table 6 summarizes drugs and metabolites that have been detected in amniotic fluid and breast milk.

3. BREAST MILK

3.1. Anatomy and Physiology

The female breast consists of 15–20 lobes of milk-secreting glands, embedded in the fatty tissue. During pregnancy, estrogen and progesterone, secreted in the ovary and placenta, cause the milk-producing glands to develop and become active. The ducts of these glands have their outlet in the nipple, and by mid-pregnancy, the mammary glands are prepared for secretion. Colostrum, a creamy white to yellow pre-milk fluid, may be expressed from the nipples during the last trimester of pregnancy. The pH of colostrum more closely resembles that of plasma. This difference in composition and pH can influence the drug content. This fluid, which is a rich source of protein, fat, carbohydrate, and antibodies, is replaced with breast milk within 3 days of delivery of the fetus and placenta. Proteins, sugars, and lipids in the milk provide initial nourishment to the newborn infant. The production of between 600 and 1000 mL of milk per day by the milk-secreting cells is stimulated by the pituitary hormone, prolactin. Contraction of the myoepithelial cells surrounding the alveoli allows the milk to be expressed into the duct system.

3.2. Drug Transfer

In addition to widespread prescription drug use during pregnancy *(9)*, more than 90% of women receive medication during the postpartum week *(21)*. This is accompanied by a near threefold increase in the number of women who breast-fed their infants in recent years. Fortunately, however, most drugs given to nursing mothers reach infants in smaller amounts compared with drugs given during pregnancy *(21)*.

The transfer of drug into the milk depends on metabolism, protein binding, and the circulation of blood in the mammary tissue. The total protein concentration of plasma (75 g/L) far exceeds that of breast milk (8 g/L), which limits the passage of highly protein-bound drugs into the milk. Passive diffusion is largely responsible for transporting the drug across the mammary epithelium, interstitial fluid, and plasma membranes into the milk. Drugs that are extensively protein bound may not readily pass into the milk, but emulsified fats contained in the milk may concentrate highly lipid-soluble drugs. Drugs with molecular weights less than 200 Da readily pass through small pores in the semi-permeable membrane. The mildly acidic pH of breast milk tends to trap weakly basic drugs.

3.3. Sample Collection and Drug Analysis

Fluid is collected using a special device such as a breast milk pump, after which well-established analytical techniques may be used to detect drugs-of-abuse (Table 5). Breast milk, which contains protein (1%), lipid (4%), lactose (7%), and water (88%), is mildly acidic. The average pH is 7.08 although it may range from 6.35–7.35. However, the high lipid content of milk may interfere or decrease the extraction efficiency or recovery of some drugs. Additional washing with non-polar solvents such as hexane may be necessary to remove excess lipids prior to chromatographic analyses. The effect of natural emulsifying agents in breast milk, which have detergent-like activity, may interfere with antibody–antigen reactions that take place in immunoassay screening tests. The daily variation of breast milk composition, combined with drug dose and time of administration relative to the expression of milk, is likely to affect the amount of drug present and the effect on the infant. The concentration of drug in the breast milk is subject to both within and between subject variation, further confounding attempts to generalize infant risk assessment. Composition is known to vary with time of day and method of sampling. The lipid content of the milk varies not only daily but also during a single feed; the latter portion of expressed milk may contain a several-fold increase in fat. Changes in pH, lipid or protein content throughout the stages of lactation and throughout the feeding interval are expected to influence the rates of drug transfer.

3.4. Toxicological Findings

Small, relatively lipophilic drugs may readily diffuse into the breast milk and become concentrated in the lipid-rich fluid. Furthermore, basic drugs may become sequestered because of ion trapping. For this reason, PCP was detected in breast milk 41 days following cessation of maternal drug use. A PCP concentration of 3 μg/L was detected almost 6 weeks after drug use *(15)*.

Therapeutic use of narcotic analgesics during the delivery and postpartum phase is not uncommon. Although transfer of morphine to the infant through the milk was once considered negligible, low oral doses of morphine to nursing mothers produced drug concentrations in milk up to 100 µg/L and were extremely variable between feeds *(22)*. The morphine concentration in the infant serum was 4 µg/L, which was considered to be in the analgesic range. Based on clearance and bioavailability, the authors suggested that the infant received between 0.8 and 12% of the maternal dose. In seven patients receiving intravenous morphine following cesarean delivery, morphine and morphine-6-glucuronide concentrations in colostrum were 0–48 and 0–1084 µg/L, respectively *(23)*. Morphine concentrations were always smaller in the milk compared to the plasma. However, morphine-6-glucuronide concentrations were always higher. Although the low bioavailability of morphine (20–30%) lessens the risk to the infant, it should also be considered that inactive conjugated metabolites such as morphine-3-glucuronide in breast milk may undergo reactivation by deconjugation in the gastrointestinal tract of the infant. Neonatal abstinence syndrome is largely associated with narcotic analgesics, and the perceived significance and risk associated with maternal opioid use is a matter of debate in the scientific literature.

In the past, breastfeeding was discouraged among women who were receiving more than 20 mg methadone a day. However, a recent study suggested that breastfeeding among methadone-maintained patients is generally acceptable and even advocates its use for the abatement of symptoms associated with neonatal abstinence syndrome *(24)*. Methadone is lipophilic and highly protein bound. Peak methadone concentrations are reported to occur about 4 h after oral administration, with an average of 2.2% of the daily dose being secreted into the milk.

Methadone maintenance (50 mg/day) in a drug-dependent nursing mother produced breast milk concentrations between 20 and 120 µg/L in the first 24 h after the dose, substantially lower than maternal plasma concentration *(20)*. Although methadone concentrations are typically lower in milk compared with maternal plasma, a concentration of 5.7 mg/L in breast milk was reported in another study *(25)*.

Other opioids, including hydromorphone and oxycodone, have also been detected in breast milk. In one study of eight women who received intranasal hydromorphone, it was estimated that the infant received approximately 0.67% of the maternal dose *(26)*. Meperidine and its active metabolite normeperidine were detected in breast milk at concentrations in the range 36–314 and 0–333 µg/L following postpartum analgesia *(27)*. Although some studies suggest the quantities of drug transferred to the infant are minimal, others have shown decreased neonatal alertness and neurobehavioral outcome compared to morphine *(28,29)*. Fentanyl was found to concentrate in lipid-rich colostrum at much higher concentration compared to maternal serum.

Benzodiazepines tend to be lipophilic and uncharged, factors that facilitate their transport across membranes into the milk. More water-soluble benzodiazepines like temazepam are less likely to accumulate in breast milk compared with lipophilic analogs like diazepam. Although a number of benzodiazepines have been detected in breast milk, concentrations are typically very much lower than those found in maternal plasma. In one report of a woman receiving treatment for benzodiazepine withdrawal, maximum concentrations of diazepam, nordiazepam and oxazepam in breast milk were 307, 141 and 30 µg/L, respectively *(30)*. Although the dose of diazepam delivered to the infant through breast milk was estimated to be 4.7% of the maternal dose, the immaturity of hepatic enzymes, slow metabolism, and elimination of some benzodiazepines in the infant should also be considered.

A number of antidepressant medications that are frequently used for postpartum depression have been reported in breast milk although concentrations are generally considered to be low *(31)*. Transfer of selective serotonin reuptake inhibitors (SSRIs) including fluoxetine, sertraline, fluvoxamine, paroxetine, and citalopram are reported to deliver 0.5–10% of the maternal dose to the infant *(32)*. Detectable amounts of parent drug in the milk at the nursing infant have been measured. Although short-term adverse effects of SSRI administration through breast milk are rare, more research on the long-term consequences of maternal drug use are needed. Transfer of fluoxetine into breast milk is not unexpected given its lipophilicity and basic nature. In a study of 10 women, the average dose of fluoxetine administered to nursing infants was estimated to be 10.8% of the maternal dose although as much as 28.6% was delivered to one infant. In this case, a fluoxetine dosing regimen of 0.27 mg/kg/day produced mean fluoxetine and norfluoxetine concentrations of 41 and 96 µg/L respectively in breast milk *(33)*. Peak concentrations in milk were achieved in approximately 6 h. Less than 10% of the therapeutic dose is often considered as a reference point of safety for nursing infants *(34)*. Citalopram and fluoxetine appear to deliver a comparable dose to nursing infants. Trough plasma concentrations of citalopram in nursing infants were 64% those of maternal plasma. Furthermore, citalopram metabolites were present in milk a two to threefold higher concentration compared with maternal plasma *(35)*. Fluvoxamine, paroxetine, and sertraline, which produce peak concentrations in breast milk at 7–10 h, are reported to deliver less drug to the infant than either fluoxetine or citalopram *(36)*.

Stimulant drugs that are basic are particularly susceptible to ion trapping and accumulation in breast milk. Following a therapeutic dosing regimen for narcolepsy, amphetamine concentrations in four milk samples were in the range

55–138 μg/L. These concentrations exceeded those found in maternal plasma by three to sevenfold *(37)*. In this report the dose of amphetamine (20 mg/day) was very much lower than those used by the drug abusing population. Animal studies using rats have also confirmed that cocaine preferentially partitions into breast milk. Cocaine and several of its metabolites have been detected in breast milk. The parent drug, which is the predominant analyte in breast milk, was as high as 12 mg/L in one study of six women who used cocaine *(38)*. Intoxication of breast-fed infants by cocaine-abusing mothers has been reported *(39)*.

Despite the fact that as many as 34% of pregnant women are reported to have used marijuana, the effect of postnatal drug exposure is not completely understood. The primary active ingredient, Δ-9-tetrahydrocannabinol (THC), is lipophilic and readily transfers into the milk where it may accumulate. Concentrations of THC are reported to be eightfold higher in breast milk compared with maternal plasma *(40)*. THC concentrations in breast milk of 105 and 340 μg/L have been reported, the latter in a chronic marijuana smoker *(41,42)*.

4. INTERPRETATION

The direct impact of specific drugs on the newborn child is difficult to evaluate. Many substance-abusing women use multiple drugs, receive inadequate health care, and may be predisposed to other health problems that may impact both neonatal and maternal outcomes. A number of illicit, prescription, and over-the-counter drugs have been detected in amniotic fluid and breast milk using well-established immunochemical, chromatographic, and spectroscopic techniques. Much of the human data to date is quite limited and predominantly consists of individual case reports. Small animal studies, although more numerous, are subject to biological scaling and possible differences in drug metabolism, distribution and toxicity.

Long-term implications of prenatal drug exposure are limited, and many consequences of fetal drug exposure are still unknown. Despite adequate understanding of the maternal consequences of drug abuse, fetal consequences for many drugs are poorly understood and this is a challenging area of maternal–fetal medicine.

Issues concerning use of prescription drugs during lactation are clinically important but complex. Much of the information available is based upon short-term or single-dose studies. The assessment of adverse drug reactions in neonates and infants are difficult to discern, and the effects of long-term drug exposure have not been fully elucidated for many drugs. The use of illicit drugs poses many of the same issues but is also compounded by other complications such as multiple drug use and health issues associated with substance-abusing individuals.

Although analytical methodology that is routinely used in the toxicology laboratory can be used to detect drugs or metabolites in specimens of maternal origin, appropriate cutoff concentrations and detection limits must be utilized. Clinical studies reported to date are somewhat limited in scope. However, additional research is needed in order to fully understand the mechanisms that influence the transfer of drugs within the maternal-fetal complex and between mother and infant.

REFERENCES

1. The NHSDA Report, *Pregnancy and Illicit Drug Use*, Office of Applied Studies (OAS), Substance Abuse and Mental Health Services Administration, Rockville, MD, 2001.
2. The NHSDA Report, *Substance Use Among Pregnant Women During 1999 and 2000*, Office of Applied Studies (OAS), Substance Abuse and Mental Health Services Administration, Rockville, MD, 2002.
3. Huestis MA, Choo RE. Drug abuse's smallest victims: *in utero* drug exposure. *J Forensic Sci.* 2002;128:20–30.
4. Ostrea EM Jr, Welch RA. Detection of prenatal drug exposure in the pregnant women and her newborn infant. *Clin Perinatol.* 1991;18(3):629–45.
5. Ostrea EM Jr, Brady M, Gause S, Raymundo AL, Stevens M. Drug screening of newborns by meconium analysis: a large-scale, prospective, epidemiologic study. *Pediatrics.* 1992;89(1):107–13.
6. Lester BM, ElSohly M, Wright LL, Smeriglio VL, Verter J, Bauer CR, Shankaran S, Bada HS, Walls HH, Huestis MA, Finnegan LP, Maza PL. The Maternal Lifestyle Study: drug use by meconium toxicology and maternal self-report. *Pediatrics.* 2001;107(2):309–17.
7. Doering PL, Boothby LA, Cheok M. Review of pregnancy labeling of prescription drugs: is the current system adequate to inform of risks? *Am J Obstet Gynecol.* 2002;187(2):333–9.
8. Malm H, Martikainen J, Klaukka T, Neuvonen PJ. Prescription of hazardous drugs during pregnancy. *Drug Saf.* 2004;27(12):899–908.
9. Andrade SE, Gurwitz JH, Davis RL, Chan KA, Finkelstein JA, Fortman K, McPhillips H, Raebel MA, Roblin D, Smith DH, Yood MU, Morse AN, Platt R. Prescription drug use in pregnancy. *Am J Obstet Gynecol.* 2004;191(2): 398–407.
10. Kwong TC, Ryan RM. Detection of intrauterine illicit drug exposure by newborn drug testing. National Academy of Clinical Biochemistry. *Clin Chem.* 1997;43(1):235–42.
11. Chiriboga CA. Fetal alcohol and drug effects. *Neurologist.* 2003;9(6):267–79.
12. Pacifici GM, Sawe J, Kager L, Rane A. Morphine glucuronidation in human fetal and adult liver. *Eur J Clin Pharmacol.* 1982;22(6):553–8.

13. Jain L, Meyer W, Moore C, Tebbett I, Gauthier D, Vidyasagar D. Detection of fetal cocaine exposure by analysis of amniotic fluid. *Obstet Gynecol.* 1993;81(5 Pt 1):787–90.
14. Apple FS, Roe SJ. Cocaine-associated fetal death *in utero. J Anal Toxicol.* 1990;14(4):259–60.
15. Kaufman KR, Petrucha RA, Pitts FN Jr, Weekes ME. PCP in amniotic fluid and breast milk: case report. *J Clin Psychiatry.* 1983;44(7):269–70.
16. Shannon C, Jauniaux E, Gulbis B, Thiry P, Sitham M, Bromley L. Placental transfer of fentanyl in early human pregnancy. *Hum Reprod.* 1998;13(8):2317–20.
17. Gerdin E, Rane A, Lindberg B. Transplacental transfer of morphine in man. *J Perinat Med.* 1990;18(4):305–12.
18. Potsch L, Skopp G, Emmerich TP, Becker J, Ogbuhui S. Report on intrauterine drug exposure during second trimester of pregnancy in a heroin-associated death. *Ther Drug Monit.* 1999;21(6):593–7.
19. Harper RG, Solish G, Feingold E, Gersten-Woolf NB, Sokal MM. Maternal ingested methadone, body fluid methadone, and the neonatal withdrawal syndrome. *Am J Obstet Gynecol.* 1977;129(4):417–24.
20. Kreek MJ, Schecter A, Gutjahr CL, Bowen D, Field F, Queenan J, Merkatz I. Analyses of methadone and other drugs in maternal and neonatal body fluids: use in evaluation of symptoms in a neonate of mother maintained on methadone. *Am J Drug Alcohol Abuse.* 1974;1(3):409–19.
21. Anderson PO. Drug use during breast-feeding. *Clin Pharm.* 1991;10(8):594–624.
22. Robieux I, Koren G, Vandenbergh H, Schneiderman J. Morphine excretion in breast milk and resultant exposure of a nursing infant. *J Toxicol Clin Toxicol.* 1990;28(3):365–70.
23. Baka NE, Bayoumeu F, Boutroy MJ, Laxenaire MC. Colostrum morphine concentrations during postcesarean intravenous patient-controlled analgesia. *Anesth Analg.* 2002 ;94(1):184–7.
24. Ballard JL. Treatment of neonatal abstinence syndrome with breast milk containing methadone. *J Perinat Neonatal Nurs.* 2002 ;15(4):76–85.
25. Blinick G, Inturrisi CE, Jerez E, Wallach RC. Methadone assays in pregnant women and progeny. *Am J Obstet Gynecol.* 1975;121(5):617–21.
26. Edwards JE, Rudy AC, Wermeling DP, Desai N, McNamara PJ. Hydromorphone transfer into breast milk after intranasal administration. *Pharmacotherapy.* 2003;23(2):153–8.
27. Quinn PG, Kuhnert BR, Kaine CJ, Syracuse CD. Measurement of meperidine and normeperidine in human breast milk by selected ion monitoring. *Biomed Environ Mass Spectrom.* 1986;13(3):133–5.
28. Borgatta L, Jenny RW, Gruss L, Ong C, Barad D. Clinical significance of methohexital, meperidine, and diazepam in breast milk. *J Clin Pharmacol.* 1997;37(3):186–92.
29. Wittels B, Glosten B, Faure EA, Moawad AH, Ismail M, Hibbard J, Senal JA, Cox SM, Blackman SC, Karl L, Thisted RA. Postcesarean analgesia with both epidural morphine and intravenous patient-controlled analgesia: neurobehavioral outcomes among nursing neonates. *Anesth Analg.* 1997;85(3):600–6.

30. Dusci LJ, Good SM, Hall RW, Ilett KF. Excretion of diazepam and its metabolites in human milk during withdrawal from combination high dose diazepam and oxazepam. *Br J Clin Pharmacol.* 1990;29(1):123–6.
31. Buist A, Norman TR, Dennerstein L. Breastfeeding and the use of psychotropic medication: a review. *J Affect Disord.* 1990;19(3):197–206.
32. Epperson N, Czarkowski KA, Ward-O'Brien D, Weiss E, Gueorguieva R, Jatlow P, Anderson GM. Maternal sertraline treatment and serotonin transport in breast-feeding mother-infant pairs. *Am J Psychiatry.* 2001;158(10):1631–7.
33. Taddio A, Ito S, Koren G. Excretion of fluoxetine and its metabolite, norfluoxetine, in human breast milk. *J Clin Pharmacol.* 1996;36(1):42–7.
34. Bennett PN and the WHO Working Group (1988). *Drugs and Human Lactation,* Amsterdam, Elsevier.
35. Heikkinen T, Ekblad U, Kero P, Ekblad S, Laine K. Citalopram in pregnancy and lactation. *Clin Pharmacol Ther.* 2002;72(2):184–91.
36. Spigset O, Carieborg L, Ohman R, Norstrom A. Excretion of citalopram in breast milk. *Br J Clin Pharmacol.* 1997;44(3):295–8.
37. Steiner E, Villen T, Hallberg M, Rane A. Amphetamine secretion in breast milk. *Eur J Clin Pharmacol.* 1984;27(1):123–4.
38. Winecker RE, Goldberger BA, Tebbett IR, Behnke M, Eyler FD, Karlix JL, Wobie K, Conlon M, Phillips D, Bertholf RL. Detection of cocaine and its metabolites in breast milk. *J Forensic Sci.* 2001;46(5):1221–3.
39. Chasnoff IJ, Lewis DE, Squires L. Cocaine intoxication in a breast-fed infant. *Pediatrics.* 1987;80(6):836–8.
40. Astley SJ, Little RE. Maternal marijuana use during lactation and infant development at one year. *Neurotoxicol Teratol.* 1990;12(2):161–8.
41. Perez-Reyes M, Wall ME. Presence of delta-9-tetrahydrocannabinol in human milk. *N Engl J Med.* 1982;307(13):819–20.
42. Reisner SH, Eisenberg NH, Stahl B, Hauser GJ. Maternal medications and breast-feeding. *Dev Pharmacol Ther.* 1983;6(5):285–304.

Chapter 2

Drugs-of-Abuse in Meconium Specimens

Christine M. Moore

Summary

Meconium is the first fecal material passed by the newborn. It begins to form between 12 and 16 weeks of gestation and therefore, may provide a history of in utero drug exposure during the second and third trimesters. Although meconium is easy to collect, small sample sizes, lack of homogeneity, different metabolic profiles, and the requirement for low limits of detection present analytical challenges for drug testing. Immunoassay screening assays and mass spectrometric-based confirmation procedures have been described for the common drugs-of-abuse.

Key Words: Meconium, forensic toxicology, drugs-of-abuse.

1. INTRODUCTION

Fetal exposure to drugs, alcohol, or other xenobiotics results in numerous adverse effects for the newborn. Maternal use of cocaine, methamphetamine (MA), and/or phencyclidine (PCP) has been consistently reported as a co-factor in births involving respiratory problems, intracranial bleeding, placental abruption, premature labor, low birth weight, and small head size babies, as well as fetal death *(1–5)*. Behavioral consequences later in childhood have also been studied. Beeghly et al. *(6)* recently showed that newborns with prenatal cocaine exposure (PCE) had lower receptive language than children not exposed to cocaine at 6 years, but that difference had modified by age 9. Age, birth

From: *Forensic Science and Medicine: Drug Testing in Alternate Biological Specimens*
Edited by: A. J. Jenkins © Humana Press, Totowa, NJ

weight, and gender seemed to moderate the relationship between PCE and the language ability of school age children.

Maternal opiate abuse can result in newborns displaying irritability, tremors, and seizures, which are often consistent with opiate withdrawal symptoms *(7,8)*. Although *cannabis* was considered a fairly harmless drug, Hurd et al. *(9)* recently reported that maternal marijuana use impaired growth in mid-gestation fetuses. Other researchers have reported adverse effects of marijuana use on the newborn *(10)* and on the adolescent offspring of mothers smoking marijuana. Porath and Fried recently reported that "maternal cigarette smoking and marijuana use during pregnancy are risk factors for later smoking and marijuana use among adolescent offspring, and add to the weight of evidence that can be used in support of programs aimed at drug use prevention and cessation among women during pregnancy" *(11)*.

As these major drug classes cause negative effects in the offspring, it is imperative that a reliable diagnosis of fetal drug exposure be made as soon as possible, in order that the appropriate care and treatment be given to both the newborn and the mother.

1.1. Acceptance of Meconium Analysis

The first reports of the use of meconium to determine fetal drug exposure were published in 1989 *(12)*, and various patents have been awarded based on methods of analysis since that time *(13–17)*. Meconium analysis has become routine in many hospitals, as it is a depository for drugs to which the fetus has been exposed during the latter half of pregnancy and provides a much longer history of fetal drug exposure than urine. Various review articles on the analysis of drugs-of-abuse in meconium have been published *(18,19)*. Several large-scale studies involving the use of meconium analysis have been performed over recent years. For example, meconium specimens from 8527 newborns were analyzed by immunoassay with gas chromatography/mass spectrometry (GC/MS) confirmation for metabolites of cocaine, opiates, cannabinoids, amphetamines (APs), and PCP as part of The National Maternal Lifestyle Study *(20)*. The prevalence of cocaine/opiate exposure was determined to be 10.7% with the majority of neonates (9.5%) exposed to cocaine. However, exposure status varied by site and was higher in low-birth-weight infants (18.6% for very low birth weight and 21.1% for low birth weight). In the cocaine/opiate-exposed group, 38% were cases in which the mother denied use, but the meconium was positive. The report concluded that accurate identification of prenatal drug exposure was improved with GC/MS confirmation of the meconium assay, and maternal interview was taken into account *(20)*. Several other researchers have shown increased detection rate of fetal drug exposure when using meconium compared to urine *(21,22)*. In 2003, Bar-Oz et al. tested paired samples of neonatal hair and meconium for cocaine, benzoylecgonine (BZE), opiates, cannabis, benzo-diazepines, methadone, and barbiturates. They reported that meconium was marginally more sensitive than neonatal hair for the detection of cocaine and marijuana. Both meconium and hair were more effective than urine for the detection of drug exposure *(23)*.

2. COMPOSITION OF MECONIUM

Meconium is the first fecal material passed by a newborn and is normally excreted 1–5 days after birth. It begins to form between the 12th and 16th week of gestation and is a cumulative deposit thereafter. Generally, it represents the intestinal contents of the fetus, providing a history of fetal swallowing and bile excretion, and can provide the physician guidance on gastrointestinal function. The color is generally dark-green because of the presence of bile pigmentation. Meconium consists predominantly of water, but other major components include

- mucopolysaccharides
- epithelial cells
- lipids and proteins
- cholesterol and sterol precursors
- blood group substances
- squamous cells
- enzymes
- bile acids and salts
- residual amniotic fluid.

Meconium is occasionally passed in utero when the fetus has reached gastrointestinal maturity late in gestation or the baby is in distress, producing meconium-stained amniotic fluid. However, in 12–25% of deliveries involving the meconium passage in utero, the cause is not known.

3. DEPOSITION OF DRUGS IN THE FETUS

Because of the obvious ethical limitation of providing known amounts of drug to pregnant women, much of the research in the area of drug deposition has been conducted in animals. Studies using pregnant sheep and guinea pigs have demonstrated the presence of both parent drug and metabolites in offspring. However, it still remains difficult to predict the metabolic fate of a drug in the maternal–fetal unit as the various breakdown pathways (oxidation, conjugation, hydrolysis reduction) can be promoted or retarded by fetal or placental enzymes. In general terms, reaction rates increase with gestational age, as maturation of the metabolic pathways occurs, but presence of a specific metabolite in the fetus may not indicate the ability of the fetus to metabolize the drug, as passive diffusion through the placenta may have occurred. Placental microsomes containing cholinesterase may be responsible for some enzymatic conversion of drugs to their metabolites as they attempt to cross the placental barrier. Placental transfer of drugs is affected by blood flow, drug ionization, and degree of protein binding of the drug to either maternal or fetal plasma.

The kinetics of drug transfer have been reviewed *(24)*, and it has been suggested that drugs reach the fetus by various complex pathways:

- Passive diffusion of small molecule lipid-soluble drugs across the placental barrier, which may be further metabolized by the fetus depending on gestational age
- Binding of drugs and/or metabolites to proteins in the amniotic fluid, which is then swallowed by the fetus. Sustained swallowing of amniotic fluid may prolong fetal exposure to drugs *(25)*.

Therefore, drugs enter the fetal circulation through various pathways that depend upon many parameters. The meconium is the final depository for drugs to which the fetus was exposed, because drugs in the bile are deposited into meconium and drugs in the urine are deposited into the amniotic fluid, which is then swallowed by the fetus. Consequently, meconium is not a homogenous specimen, because it is produced in "layers" as excreted products are stored. This is the main disadvantage to testing meconium for drugs-of-abuse, because frequently not all the specimen is collected.

Because of the numerous variables involved over a period of 20 weeks (latter half of pregnancy), such as frequency and nature of drug abuse, nutrition, smoking, other diseases, and general maternal health, the metabolic profile of a drug appearing in the newborn is not the same as in a normal adult. This observation has also contributed to the difficulty of meconium analysis to avoid false-negative results.

4. SAMPLE PREPARATION AND INSTRUMENTAL TESTING METHODOLOGIES

4.1. Immunochemical Screening Assays

Meconium is a complex matrix containing waste products and pigments. The amount of drugs and metabolites generally detected in meconium are much lower than concentrations present in urine samples. There are numerous publications regarding the screening of meconium specimens. In the original patents *(13)*, meconium (0.5 g) was taken directly from the diaper of the newborn. The sample was mixed with distilled water (10 mL) and concentrated hydrochloric acid (1 mL). This homogenate was filtered through glass wool, the filtrate centrifuged, and the supernatant tested for morphine and BZE using

Abuscreen Radioimmunoassay (RIA). For cannabinoids, methanol (0.4 mL) was added to the sample (0.1g). The specimen was mixed, allowed to stand at room temperature, and then centrifuged. An aliquot of the supernatant was tested by RIA. The cut-off values reported were 15, 25, and 50 ng/mL for cocaine, morphine, and cannabinoids, respectively.

Since this original work, many extraction procedures have been reported. In general, the meconium is homogenized in an organic solvent, such as methanol or acetonitrile and/or buffer, to help the drugs distribute throughout the matrix. The concept is to solubilize the drugs in the sample so they can be tested using standard screening formats. The specimen is centrifuged, and the supernatant is analyzed using RIA *(26,27)*, enzyme multiplied immunoassay technique (EMIT) *(28,29)*, enzyme-linked immunosorbent assay (ELISA), or fluorescence polarization immunoassay (FPIA) *(30,31)*. However, it is difficult to achieve the detection levels required for valid analysis without some form of extraction of drugs from the matrix because of the interference of the pigmentation and turbidity of the extract, especially when enzyme or fluorescent detection is used. Chen and Raisys *(28)* analyzed the drugs in meconium by homogenization of the sample in methanol, followed by solid-phase extraction (SPE), before screening with EMIT®. Moriya et al. *(29)* extracted drugs from meconium with chloroform–isopropanol (3:1) and screened by EMIT. The reported detection limits for BZE, D-MA, morphine, and PCP were 250, 730, 110, and 100 ng/g, respectively, which even with the extraction are relatively high.

ElSohly et al. *(32)* described the development and validation of methods for meconium sample preparation for both screening by EMIT and TDx and confirmation by gas chromatography–mass spectrometry (GC–MS) of meconium extracts for cannabinoids, cocaine, opiates, APs, and PCP. The procedures necessitated specimen clean-up prior to screening to achieve the required detection levels. Cut-off levels were administratively set at 20 ng/g for 11-nor-Δ9-tetrahydrocannabinol-9-COOH (THC-COOH) and PCP and at 200 ng/g for BZE, morphine, and APs.

4.1.1. FALSE-POSITIVE AND FALSE-NEGATIVE RESULTS IN MECONIUM SCREENING ASSAYS

Several research groups have warned against relying on positive meconium results using immunoassay only, without confirmation *(20,33,34)*. It is necessary for valid, relevant, and reliable detection levels to be observed within the laboratory. Possible reasons for false-negative results include:

• insufficient sensitivity or incorrect selection of the target drug
• drug metabolites or structurally related compounds contributing to immunoassay response, which are then not part of the confirmatory profile

- inadequate sample clean-up prior to analysis
- lack of cleanliness of the extract resulting in interference in the assay.

Conversely, false-positive results can be observed due to

- increased sensitivity of the immunoassay systems
- unrelated compounds contributing to immunoassay response
- reliance on screen only results.

ElSohly et al. reported the analysis of 30 meconium specimens by GC–MS analysis for all analytes regardless of the screening results to determine the false-negative rate, if any, of the immunoassay. They determined that there were no false negatives detected using their immunoassay procedures, but the confirmation rate for the positive specimens was generally low. Surprisingly, no specimens were positive for APs. The lowest rate of confirmed positives was found with the cannabinoids. The authors suggested that THC metabolites other than free THC-COOH may be major contributors to the immunoassay response in meconium *(32)*.

4.2. Confirmatory Assays

As previously mentioned, presumptively positive immunoassay tests require that a second assay based on a scientifically separate principle be used to ensure the identification of the drug *(20,33)*. For several years, GC–MS was the preferred confirmatory procedure for many drugs-of-abuse. High-pressure liquid chromatography (LC) with ultraviolet, fluorescence, or diode array detection procedures have been reported *(31,35–37)*, but advancing technology has allowed the use of LC coupled to MS to be applied to the analysis of meconium *(38)*.

4.2.1. COCAINE

The majority of research has focused on cocaine, as it is considered to be a widely abused drug and there are many reports of neonatal consequences of maternal use. At least fifteen metabolites or adducts of cocaine plus alcohol, as well as cocaine itself, are reportedly present in meconium. These include BZE, benzoylnorecgonine (BN), norcocaine (NC), ecgonine (ECG), ecgonine methyl ester (EME), ecgonine ethyl ester (EEE), cocaethylene (CE), norcocaethylene, *meta*-hydroxybenzoylecgonine (*m*-OH-BZE), *para*- hydroxybenzoylecgonine (*p*-OH-BZE), *para*-hydroxycocaine (*p*-OH-COC), *m*-hydroxycocaine (*m*-OH-COC), anhydroecgonine methyl ester (AEME), anhydroecgonine, and cocaine-*N*-oxide.

4.2.1.1. Gas Chromatography–Mass Spectrometry

In urine, the major metabolite detected following cocaine ingestion is BZE. Therefore, most immunoassay platforms and, in earlier research confirmatory assays were targeted to the detection of BZE. The first confirmatory procedure was reported by Clark et al. in 1992 *(39)* and employed a methanolic extraction, followed by the addition of phosphate buffer, SPE, and derivatization of BZE using trimethylsilyl derivatives. The assay was linear to 10 ng/g for both cocaine and BZE. However, in 1993, Steele et al. *(40)* first identified m-OH-BZE as a major contributor to immunoassay-positive results in samples collected from cocaine-exposed newborns. Using the standard confirmation profile of cocaine and BZE, these specimens, which had screened positively, did not confirm. This observation caused the development of new procedures for a wide range of cocaine metabolites. In 1994, Lewis et al. subsequently confirmed this observation and reported that up to 23% of newborns screening positively for cocaine metabolites were not being correctly diagnosed as the confirmation profile did not include m-OH-BZE *(41)*.

Oyler et al. *(42)* conducted an extensive study into the metabolism of cocaine in the newborn following fetal exposure by analyzing meconium and urine for all the metabolites named above, except cocaine-*N*-oxide. They were the first to identify *p*-OH-BZE in meconium and suggested that this newly identified metabolite, like *m*-OH-BZE, might serve as a valuable marker of fetal cocaine exposure during pregnancy. The presence of cocaine and AEME in meconium was attributed to transfer across the placenta from the mother. However, the origin of the hydrolytic and oxidative metabolites of cocaine could not be established because they were also identified in urine specimens of adult female cocaine users and could have arisen in meconium from either fetal or maternal metabolism. Later, ElSohly et al. *(43)* reported the presence of *m*-OH-BZE and also identified *p*-OH-BZE as a cocaine metabolite in meconium. Abusada et al. *(44)* reported the presence of cocaine, BZE, EME, and CE in meconium. The presence of CE in meconium was the first identification of this important adduct, as its presence indicated maternal intake of alcohol as well as cocaine. Subsequently, CE has been included in many confirmation profiles for cocaine and its metabolites. CE was identified in 31.6% of samples containing cocaine and/or BZE *(45)*. The authors also reported that "cocaethylene accumulates in greater concentrations in meconium than urine, and is a useful analyte for identifying fetal alcohol exposure." *(45)*.

4.2.1.2. Liquid Chromatography and Liquid Chromatography–Mass Spectrometry

In 1994, Browne et al. *(35)* used high-performance liquid chromatography (HPLC) and GC/MS to analyze meconium for cocaine, NC, and CE in the meconium of premature infants. The report was the first identification of NC in meconium samples, and the protocol was adopted by Dusick et al. *(2)*, who tested the meconium of 323 very low-birth-weight newborns.

In 1993, Murphey et al. *(36)* were the first to report the presence of benzoylenorecgonine (BN) in meconium using HPLC. Following SPE of the drugs from the sample, separation of cocaine, BZE, NC, and BN was achieved using a Microsorb C18 column (100 × 4.6 mm; 3 μm particle size) with a mobile phase of 0.01M NaH_2PO_4 at pH 2.0 with 58 μm of tetrabutylammonium hydroxide and 13% acetonitrile pumped at a flow rate of 1 mL/min. The assay was linear from 50 to 5000 ng/g. BN, a previously unreported metabolite, was detected in 7 of 11 meconium samples from neonates born to cocaine using women.

However, in an interesting article published in 2000, Xia et al. suggested that ECG was in fact the best marker for the determination of cocaine exposure although the authors recommended an eight-metabolite profile for confirmation. Their procedure employed SPE of the drugs followed by analysis using LC with tandem mass spectrometric detection. The HP 1100 LC system was interfaced to a Micromass Quattro II triple quadrupole MS. The authors reported the presence of ECG, *p*-OH-cocaine, and CE in 21 of 22 samples (one sample was negative for all metabolites). The presence of CE in every sample is surprising, as it is only formed when there is concurrent intake of cocaine and ethanol. This indicated that all the women in the study also exposed their newborns to alcohol. The authors attempted to explain this, stating CE was only present above the level of 10 ng/g in 15 of the 21 samples (71.4%). The detection of ECG in all the specimens is particularly interesting, as the authors note that specific washes of the solid-phase cartridge (hydrochloric acid, followed by extensive methanol washing) as recommended in the manufacturer's procedure would in fact remove all the ECG prior to analysis. Along with the absence of derivatization needed for GC/MS determination, this may account for the reason ECG has previously not been considered a major metabolite in meconium. ECG was measured at the highest median concentration in all the samples, whereas the median concentration of the *meta*-hydroxylated cocaine and BZE were higher than the corresponding *para*-hydroxylated metabolites. However, the *para*-hydroxylated metabolites were present in more specimens than the *meta*-hydroxy cocaine and BZE. The authors detected all the metabolites in many of the samples. They recommended a profile of eight analytes: cocaine, CE, ECG, *m*-OH-BZE, *p*-OH-BZE, *p*-OH-cocaine, NC, and BZE be monitored in meconium analysis to provide the greatest utility in the detection of fetal exposure to cocaine using meconium specimens *(46)*. Cocaine-*N*-oxide, a metabolite that had previously only been identified in a single case study *(47)*, was detected in 12 of the 21 samples (57%).

In 2005, Pichini et al. applied a previously reported LC/MS procedure to the determination of *m*- and *p*-OH-BZE in meconium using nalorphine as internal standard. The method included a methanol extraction, evaporation of the solvent, and sample preparation using solid-phase columns. The drug recovery was 60–65%. Reversed-phase chromatography with a C8 analytical column (150 mm × 4.6 mm) was used, with a gradient of 1% acetic acid–acetonitrile as the mobile phase. The system was coupled to atmospheric pressure ionization electrospray-MS single-ion monitoring mode. This method was valid from 5 to 1000 ng/g of meconium and when applied to authentic specimens produced quantitative values between 7 and 338 ng/g in meconium for *m*-OH-BZE and values between 7 and 319 ng/g for *p*-OH-BZE *(48)*. In all positive specimens, cocaine and/or BZE were also present in contrast to other reports *(41,43)*. Even though *m*-OH-BZE is considered to be glucuronide bound in meconium, there is sufficient free drug to be detected in a standard confirmation profile using GC/MS without the need for hydrolysis.

4.2.2. OPIOIDS

4.2.2.1. Heroin, Codeine, and Metabolites

4.2.2.1.1. Gas Chromatography–Mass Spectrometry. To date, there are no reports of heroin detected in meconium. In 2001, Salem et al. reported a comparison of extraction procedures for the determination of 6-acetylmorphine (6-AM), a primary metabolite of heroin, in specimens from neonates *(49)*. SPE cartridges were evaluated for their effectiveness in sample preparation. Four different types of commercially available extraction cartridges were used. 6-AM, morphine, and codeine were extracted from meconium samples using these SPE cartridges and analyzed by GC–MS. The limits of quantitation (LOQ) were 20 ng/g for codeine, 10 ng/g for morphine, and 5 ng/g for 6-AM, and the assays were linear over wide concentration ranges. Using nine previously screened opiate-positive meconium specimens, the authors reported the presence of 6-AM in eight of the samples at concentrations ranging from 3.19 to 70.2 ng/g. All of the specimens also contained codeine (31.25–692.71 ng/g) and morphine (264.1–1617.09 ng/g). The detection of 6-AM, however, is not widespread and is rarely reported. In the Maternal Lifestyle Study previously described *(20)*, only one sample out of 8527 meconium samples collected was positive for 6-AM using this procedure from 182 positive opiate screens (0.54%). In this study, a screening cut off of 50 ng/g was used, and this may be too high for consistent confirmation. Morphine accounted for the majority of the positive results, detected in 101 specimens (55.5%), codeine in 83 (45.6%), hydrocodone in one sample (0.54%), and hydromorphone in three samples (1.7%). Morphine, a metabolite of both heroin and codeine, is much more commonly encountered in meconium analysis *(20,32,50)* and has been reported in the meconium from stillborn babies *(51)*.

4.2.2.1.2. Liquid Chromatography and Liquid Chromatography–Mass Spectrometry. The application of HPLC, LC/MS, and LC/MS/MS to opiates in meconium has not been widely reported; however, the ability to analyze glucuronide-bound drugs without hydrolysis is a distinct advantage to these techniques. Pichini et al. *(38)* were the first to report a procedure based on LC–MS for the simultaneous determination of 6-AM, morphine, morphine-6-glucuronide, morphine-3-glucuronide, codeine, cocaine, benzoylecgonine, and CE in meconium using nalorphine as internal standard. The analytes were extracted from meconium using methanol for 6-AM, morphine, codeine, cocaine, BZE, and CE. Following solubilization, the supernatant was evaporated and part of the specimen reconstituted in phoshate buffer before being subjected to further sample clean-up using mixed mode SPE columns for 6-AM, morphine, codeine, cocaine, BZE, and CE. For the glucuronides, a portion of the methanol extract was reconstituted in 0.01 M ammonium hydrogen carbonate buffer (pH 9.3) and subjected to SPE using ethyl columns. The analytical column was a C8 reversed-phase column using a gradient of acetic acid 1%-acetonitrile as a mobile phase. Analytes were determined in LC-MS single ion monitoring mode with atmospheric pressure–electrospray (ESI) interface. The linearity of the procedure was 5–1000 ng/g. When applied to patient samples, the presence of 6-AM was confirmed, an observation that had previously only been reported by Salem et al. *(49)*, at concentrations of 5, 6, and 142 ng/g in meconium. The concentrations of 6-AM measured in meconium appear to be low and require sensitive technology to identify fetal exposure to heroin. Pichini et al. were also the first to identify morphine-3- and morphine-6-glucuronide in meconium, with one specimen containing 120 ng/g of morphine-3-glucuronide and 91 ng/g of morphine-6-glucuronide. An important practical advantage of using LC for the chromatographic separation of opiates is that hydrolysis of the specimen is not required.

4.2.2.2. Hydrocodone and Hydromorphone

4.2.2.2.1. Gas Chromatography–Mass Spectrometry. In 1995, Moore et al. reported the detection of hydrocodone and hydromorphone, for the first time, as well as codeine and morphine in meconium *(52)*. Following hydrolysis of the synthetic opiates in 2.4 M hydrochloric acid (110 °C/h), 11.8 M potassium hydroxide and buffer salts were added, and the drugs were extracted using tert-butyl methyl ether. Following reacidification and back extraction, the final solvent was evaporated to dryness and the drugs derivatized with BSTFA for analysis using GC/MS. Hydrolysis of the specimens resulted in a substantial increase in the amount of codeine, hydrocodone, and hydromorphone detected but no significant increase in the morphine concentration. With the exception of morphine, the other opiates appear to be significantly glucuronide bound, and in routine procedures, hydrolysis of meconium specimens is not always conducted *(48,53)*. Habrdova et al. reported the presence of hydrocodone, norhydrocodone, and dihydrocodeine in the urine and meconium of a newborn in 2001, but few methodological details were provided. This report is the first identifying dihydrocodeine and/or norhydrocodone in meconium *(54)*.

4.2.2.3. Oxycodone

4.2.2.3.1. Gas Chromatography–Mass Spectrometry. The detection of oxycodone in meconium was reported for the first time in 2005 *(55)*. The availability of a sustained release formulation (OxyContin®) in recent years and the potential for use during pregnancy resulting in withdrawal symptoms in the newborn *(56)* prompted the development of procedures for its analysis in meconium samples. To date, there are no reports of kinetic disposition of oxycodone in meconium. For analysis, the specimens were screened with an oxycodone Direct ELISA kit, then confirmed by GC/MS. The specimen was homogenized in methanol, centrifuged, the supernatant was evaporated to dryness, and refrigerated overnight. The following day, 0.1 M hydrochloric acid was added with methoxyamine hydrochloride. The mix was incubated at room temperature, phosphate buffer added, and the specimen subjected to mixed mode solid phase extraction. The final extract was derivatized with BSTFA +1% TMCS. The authors reported positive results for three specimens, with concentrations of 62, 224, and 490 ng/g oxycodone. No metabolites were reported.

4.2.2.4. Methadone

4.2.2.4.1. Gas Chromatography–Mass Spectrometry. In 2001, ElSohly et al. described an EMIT®-ETS d.a.u. immunoassay screening method for methadone in meconium followed by GC–MS confirmation for methadone and metabolites 2-ethylidene-1,5-dimethyl-3,3-diphenylpyrrolidine (EDDP) and 2-ethyl-5-methyl-3,3-diphenylpyrroline (EMDP) *(57)*. The GC–MS method was linear between 25 and 2000 ng/g with reported LOQ of 25 ng/g for all drugs. Fifty meconium samples were screened using a cut off of 200 ng/g, and all samples screened negative. GC/MS analysis showed that four samples contained methadone (35.2–79.9 ng/g), EDDP (28.5–557.2 ng/g), or both, with no detectable amount of EMDP. The authors stated that "the negative results on the four specimens at the cut off used may be explained by the fact that EMIT-ETS d.a.u. antibody for methadone was specific to the parent drug. The results point to the fact that immunoassays should be directed to EDDP for detection of prenatal exposure of methadone through analysis of meconium specimens" *(57)*.

4.2.2.4.2. Liquid Chromatography and Liquid Chromatography–Mass Spectrometry. In 1997, Stolk et al. *(37)* first developed methods for the analysis of methadone and its principal metabolite, EDDP, in meconium based on FPIA and HPLC with diode array detection. Meconium and urine samples of 16 neonates from 15 methadone-using mothers were analyzed. Methadone, EDDP, or both were detected in both urine and meconium samples from 15 of the newborns. However, the authors reported that the amount of EDDP in meconium was approximately 10 times higher than the amount of methadone.

In 2005, Choo et al. reported a validated LC atmospheric pressure chemical ionization tandem mass spectrometry (LC–APCI–MS/MS) method for the quantification of methadone, EDDP, EMDP, and methadol in meconium. The procedure utilized sample clean-up with solid phase extraction before analysis. The LOQ was 5 ng/g for methadone, EDDP, and EMDP; 25 ng/g for methadol, and the assay was linear from 5 to 500 ng/g. This was the first report of the presence of methadol in meconium and the first LC/MS procedure for all the analytes. The authors reported the analysis of a meconium specimen collected from an infant whose mother was maintained on methadone for 19 weeks during gestation. The meconium contained 2492 ng/g of methadone, 13,188 ng/g of EDDP, and 27 ng/g of EMDP *(58)*. These results agreed with other publications, indicating the amount of EDDP in meconium is significantly higher than the level of methadone.

4.2.3. AMPHETAMINES

4.2.3.1. Gas Chromatography–Mass Spectrometry

There are few publications regarding the analysis of APs in meconium. In the large-scale Maternal Lifestyle Study (8527 meconium samples), no APs were reported to be present, even from geographic areas of high MA abuse. The lack of positive specimens in this population may have been methodologically related, as APs are notoriously volatile and may be lost during analysis. Another possibility is that, similar to cocaine and THC, there were metabolites of the APs contributing to the immunoreactive response that were not included in the confirmatory assay. Habrdova et al. reported the presence of MA and ephedrine in the urine and meconium of a newborn in 2001, but extraction recovery for the drugs was low *(54)*.

4.2.3.2. Liquid Chromatography and Liquid Chromatography–Mass Spectrometry

In 1994, Franssen et al. *(31)* included the determination of AP but not MA in the HPLC assay previously described in the opiate section of this chapter. The detection level was high (500 ng/g) and no positives were reported. More recently, Pichini et al. published a procedure based on LC–MS for the deter-mination of AP, MA, 3,4-methylenedioxymethamphetamine (MDMA), 3,4-methylenedioxyamphetamine (MDA), 4-hydroxy-3-methoxymethamphetamine (HMMA), 3,4-methylenedioxyethylamphetamine (MDEA), and *N*-methyl-1-(3,4-methylenedioxyphenyl)-2-butamine (MBDB) in meconium, using 3,4-methylendioxypropylamphetamine (MDPA) as internal standard *(58)*. The analytes were initially extracted from the matrix using a methanol–hydrochloric acid mix, followed by SPE. Chromatography was performed on a C18 reversed-phase column using a linear gradient of 10 mM ammonium bicarbonate, pH 9.0-methanol as a mobile phase. Analytes were determined in LC–MS single-ion monitoring mode with an atmospheric pressure ESI interface. The method was validated in the range 5–1000 ng/g using 1 g of meconium per assay. Mean recoveries ranged between 80 and 90% for different analytes, with only AP demonstrating a lower extraction recovery of 60%. The quantification limits were 5 ng/g meconium for AP, MA, and HMMA; 4 ng/g for MDA, MDMA, MDEA, and MBDB. The method was applied to analysis of meconium in newborns to assess fetal exposure to AP derivatives. Over 600 specimens were analyzed, and none were positive for AP, MA, HMMA, or MDA. Only one was positive for MDMA, at a concentration of 12 ng/g, the first report of such a finding. When the procedure was applied to previously analyzed samples that had been in storage at –20 °C for at least 1 year, the concentrations of both AP and MA showed excellent correlation, indicating the stability of both AP and MA in meconium. While not specifically reporting the presence of MDMA in meconium, Ho et al. *(60)* discussed the characteristics of pregnant women using Ecstasy. They concluded that pregnant women using MDMA tended to be younger, single, had more abortions, and reported higher levels of psycho-logical problems than those not using the drug. Higher rates of unplanned pregnancy, heavy alcohol, cigarette, and illicit drug use were also reported, which may cause more problems in the newborn.

4.2.4. *Phencyclidine*

With regard to reports of drugs-of-abuse and meconium, the least number of publications have addressed the analysis of PCP and/or its metabolites although some reports are available *(29)*. PCP remains a widely used illicit drug, especially among adolescents and young adults. The effects in the newborn of PCP are similar to those of cocaine. A study by Tabor et al. compared infants exposed to PCP in utero with those exposed to cocaine. Both groups had a high incidence of intrauterine growth retardation, precipitate labor, and symptoms of neonatal drug withdrawal. However, interestingly, PCP-exposed newborns were more likely to have meconium-stained amniotic fluid, and less likely to be born prematurely as cocaine-exposed infants. Based on this data, it is possible that PCP-exposed babies are not tested as often as those exposed to other drugs, as prematurity is often a risk factor for testing and meconium is less likely to be available *(61)*.

4.2.4.1. Gas Chromatography–Mass Spectrometry

In 1996, Moore et al. reported a confirmatory procedure for the determination of PCP in meconium using the selected ion storage (SIS) functions of an ion trap mass spectrometer. The method was particularly sensitive, with the authors reporting linearity up to 250 ng/g and a detection level of 5 ng/g. The procedure was useful for meconium analysis, as sample size is often limited and sensitivity is an important factor *(62)*.

4.2.5. CANNABINOIDS

Of the drugs discussed, marijuana (THC) and its metabolites are the most difficult to analyze in meconium. Wingert et al. *(50)* reported difficulty in confirming screen positive results, probably due to the low concentration present in the sample. Additionally, the major urinary metabolite of THC, 11-nor-Δ9-THC-9-carboxylic acid (THC-COOH), is glucuronide bound in meconium, requiring hydrolysis of the sample to release the drug to permit significant drug levels to be detected. Surprisingly, there are no literature reports of the detection of THC-COOH in the urine of newborn babies, so the analysis of meconium may offer a distinct advantage for the confirmation of marijuana exposure in the neonate.

4.2.5.1. Gas Chromatography–Mass Spectrometry

In 1996, Moore et al. first published a confirmation procedure for the determination of 11-nor-Δ9-tetra-hydrocannabinol-9-carboxylic acid (THC-COOH), a major metabolite of Δ9-THC in meconium *(63)*. As THC-COOH is significantly glucuronide bound in meconium, the specimen was hydrolyzed with potassium hydroxide and extracted with hexane-ethyl acetate (9:1 v,v). Following back extraction, the final extract was derivatized with MTBSTFA for GC/MS analysis. The confirmation rate at 25 ng/g for a screening concentration was 80%, using 2 ng/g as the limit of detection. The authors suggested that other metabolites of THC may be contributing to the immunoreactive response and were not part of the confirmation profile. This observation was subsequently investigated by other research groups. Feng et al. *(64)* analyzed the presence of other metabolites in meconium by developing the use of an immunoaffinity extraction column for application to various biological matrices and applying the method to meconium extraction. Using the affinity resin prepared by immobilization of THC antibody onto cyanogen bromide-activated Sepharose

4B, Δ9-THC and its major metabolites including THC-COOH, 11-hydroxy-Δ9-THC (11-OH-THC), and 8β,11-dihydroxy-Δ9-THC were extracted simultaneously from plasma or urine after enzyme hydrolysis. The modified procedure for meconium produced lower extraction efficiencies, ranging from 52 to 72% at the 10 ng/g level. The assay was linear to 100 ng/g, and the limit of detection for the metabolites ranged from 1 to 2.5 ng/g of meconium *(64)*. Analysis of 24 meconium specimens showed that 11-OH-THC was in fact an important metabolite in meconium. Having also identified a low confirmation rate for cannabinoids when only THC-COOH was considered, ElSohly et al. *(65)* investigated the possible contribution of other metabolites of THC, including glucuronides, to the overall response of the immunoassay. Δ9-THC-glucuronide was synthesized, and procedures were developed for the extraction and GC–MS analysis of THC, 11-OH-THC, 8α-and 8β-OH-Δ9-THC, 8 β,11-diOH-Δ9-THC, and THC-COOH after enzymatic hydrolysis of meconium extracts. The authors concluded that enzymatic hydrolysis of meconium extracts was necessary for efficient recovery of THC metabolites. No significant amounts of THC or its 8-OH metabolite were detected in meconium; however, 11-OH-THC and 8 β,11-diOH-Δ9-THC did show significant contribution to immunoassay response. The authors reported several specimens for which THC-COOH was not present, but 11-OH-THC and/or 8β,11-diOH-Δ9-THC were detected.

4.2.5.2. Liquid Chromatography

Only one report of the identification of THC-COOH in meconium using LC with diode array detection is available. The report is a letter to the Editor published in 1995 which provided little analytical information *(66)*.

5. INTERPRETATION ISSUES

5.1. Positive Findings

Because of the considerable number of variables during the latter half of pregnancy (~20 weeks) and the ethical limitations of providing drugs to pregnant women, all attempts at interpreting drug levels in meconium are generally based on anecdotal evidence. The presence of drugs and metabolites in neonatal meconium is scientifically defensible as an indication of drug use by the mother at some point in the latter half of pregnancy. This is provided the testing has been performed with validated laboratory testing methods, and any screening results have been confirmed by a second method based on a scientifically different principle. However, there is insufficient evidence at this time to support the interpretation of quantitative values detected in meconium.

Lester et al. in the Maternal Lifestyle Study cautioned against the use of quantitative values to determine the degree of drug exposure *(20)*.

5.2. Negative Findings

The presence of no drugs or metabolites in meconium in cases where drug use is either admitted by the mother or suspected by health professionals should be interpreted with caution. First, an insufficient specimen quantity may be the problem with premature babies or from newborns suffering from meconium-stained amniotic fluid. Second, the detection levels employed by the laboratory should be evaluated, as high cut-off concentrations will cause low-level positive specimens to be missed. Lastly, the metabolites being tested for should be assessed, because as explained earlier in the chapter, neonatal drug profiles are different than those of adults, and specific metabolites must be included in the testing profile to ensure reliability of results.

6. ADVANTAGES OF MECONIUM ANALYSIS

There are several advantages to the use of meconium as a sample matrix:

1. The major advantage of using meconium for the determination of fetal drug exposure is its ability to indicate a longer history of potential exposure than urine. Meconium can provide up to a 20 week "window" of detection assuming that meconium starts to form around the 16th week of gestation. Maternal and/or fetal urine will only indicate the presence of drugs if they were ingested within a few days of sample collection.
2. Meconium, in comparison to urine, is easy to collect. For newborn urine collection, bags are taped to the babies, and these often become detached or urine is spilled. Meconium is simply scraped from a diaper and placed in the collection vial.
3. Meconium is not an invasive sample to collect in contrast to blood or newborn saliva. As it is a normal biological waste product, it is collected rather than discarded.
4. Drugs are stable in meconium at room temperature for up to 2 weeks, which is an important advantage for routine collections at large hospitals. Immediate refrigeration after sample collection is not always possible, and if sending to a testing laboratory, shipping on dry ice increases costs.

7. DISADVANTAGES OF MECONIUM ANALYSIS

There are several disadvantages associated with meconium analysis:

1. As meconium is not a homogenous sample, it is sometimes difficult to collect all neonatal discharges, which may be days apart. The entire specimen collected should be mixed before testing to allow the drugs to distribute throughout the matrix.

2. Meconium analysis is more labor intensive and requires more time to analyze than urine. Many hospitals are not equipped to analyze meconium, so specimens are sent out to other testing facilities, resulting in increased costs and turnaround time.
3. There are differences in metabolic profiles between meconium and urine. Therefore, standard immunoassay procedures targeted toward urinalysis may

Table 1
Metabolic Profiles in Meconium

Drugs ingested	Predominant metabolites in meconium (reference)
Cocaine	Cocaine *(20–23,28,32,34,38,39,42,43,46)*
	Benzoylecgonine *(20,28,32,38,39,42,43,46)*
	Benzoylnorecgonine *(36,42,43,46)*
	Norcocaine *(34,42,43,46)*
	Cocaethylene *(20,34,38,42–44,46)*
	Norcocaethylene *(42,46)*
	m-hydroxybenzoylecgonine *(20,40–43,46,48)*
	p-hydroxybenzoylecgonine *(42,43,46,48)*
	m-hydroxycocaine *(42,43,46)*
	p-hydroxycocaine *(42,43,46)*
	Ecgonine *(42,46)*
	Anhydroecgonine *(42,46)*
	Anhydroecgonine methyl ester *(42,46)*
	Ecgonine methyl ester *(42,43,46)*
	Ecgonine ethyl ester *(42,46)*
	Cocaine-*N*-oxide *(46,47)*
Opioids	
Heroin	6-acetylmorphine *(20,38,49)* and morphine *(20,38,50–53)*
Codeine	Codeine and morphine *(20,38,52)*
Morphine	Morphine-6-glucuronide and morphine-3-glucuronide *(38)*
Hydrocodone	Hydrocodone *(20,52)*, hydromorphone *(20,52)*, and norhydrocodone *(54)*
Oxycodone	Oxycodone *(55)*
Methadone	Methadone, EDDP, EMDP *(37,57,58)*, and methadol *(58)*
Amphetamines	
Methamphetamine	Methamphetamine and amphetamine *(59)*
Amphetamine	Not known
MDMA	MDMA *(59)*
Phencyclidine	Phencyclidine *(62)*
THC	THC-COOH *(63–66)*
	11-hydroxy-THC *(64,65)*, 8-β,11-diOH-Δ9-THC *(64,65)*

cause false-negative results. Similarly, even if samples screen positively, confirmation methods may not include the appropriate metabolites *(34,41,65)*.

4. Often the sample size obtained is small, and the concentration of drugs found in meconium is relatively low. However, advances in technology have improved the sensitivity of testing procedures, so this is increasingly of less importance.

5. There are no standardized procedures, testing profiles, proficiency programs, or control specimens available to ensure laboratory quality regarding meconium analysis.

8. SUMMARY

The detection of drugs, metabolites, or markers of drug intake in the meconium of newborns has been widely reported. This chapter has discussed the major drug classes associated with maternal abuse. Table 1 summarizes the predominant analytes present in meconium for these drugs. However, two other major xenobiotics that have been investigated regarding effects on the newborn are nicotine and alcohol. The detection markers in meconium for the identification of nicotine *(67–69)*, alcohol *(70–73)*, or both *(74–76)* have been reported.

The development of faster, more sensitive procedures for the determination of drugs-of-abuse in meconium is an on-going process. The science of meconium analysis is constantly improving as advances in analytical instrumentation and technology are incorporated into hospitals and testing laboratories.

REFERENCES

1. Chasnoff IJ, Hunt CE, Kletter R, Kaplan D. Prenatal cocaine exposure is associated with respiratory pattern abnormalities. *Am J Dis Child* **143**:583; 1989.
2. Dusick A, Covert R, Schreiber M, Yee G, Browne S, Moore C, Tebbett I. Risk of intracranial hemorrhage and other adverse outcomes after cocaine exposure in a cohort of 323 very low birth weight babies. *J Pediatr* **122**:438–445; 1993.
3. Little BB, Snell LM, Gilstrap LC. Methamphetamine abuse during pregnancy: outcome and fetal effects. *Obstet Gynecol* **72**:581; 1989.
4. Meeker JE, Reynolds PC. Fetal and newborn death associated with maternal cocaine use. *J Anal Toxicol* **14**:370; 1990.
5. Little BB, Snell LM, Kleen VR. Cocaine abuse during pregnancy: maternal and fetal implications. *Obstet Gynecol* **73**:157; 1989.
6. Beeghly M, Martin B, Rose-Jacobs R, Cabral H, Heeren T, Augustyn M, Bellinger D, Frank DA. Prenatal cocaine exposure and children's language functioning at 6 and 9.5 years: moderating effects of child age, birth weight, and gender. *J Pediatr Psychol* **281**; 2005.
7. Fulroth R, Phillips B, Durand DJ. Perinatal outcome of infants exposed to cocaine and/or heroin in utero. *Am J Dis Child* **143**:905; 1989.

8. Franck L, Vilardi J. Assessment and management of opioid withdrawal in ill neonates. *Neonatal Netw* **34**:424; 1995.
9. Hurd YL, Wang X, Anderson V, Beck O, Minkoff H, Dow-Edwards D. Marijuana impairs growth in mid-gestation fetuses. *Neurotoxicol Teratol* **27**(2): 221–229: 2005.
10. Cornelius MD, Taylor PM, Geva D, Day NL. Prenatal tobacco and marijuana use among adolescents: effects on offspring gestational age, growth, and morphology. *Pediatrics* **95**(5):738–743; 1995.
11. Porath AJ, Fried PA. Effects of prenatal cigarette and marijuana exposure on drug use among offspring. *Neurotoxicol Teratol* **27**(2):267–277; 2005.
12. Ostrea EM, Brady MJ, Parks PM, Asensio DC, Naluz A. Drug screening of meconium in infants of drug-dependent mothers: an alternative to urine testing. *J Pediatr* **115**(3):474–477; 1989.
13. Ostrea EM. *Method for Determining Maternally Transferred Drug Metabolites in Newborn Infants.* U.S. Patent 5,015,589. May 14, 1991.
14. Ostrea EM. *Method for Determining Maternally Transferred Drug Metabolites in Newborn Infants.* U.S. Patent 5,185,267. February 9, 1993.
15. Lewis DE. *Forensically Acceptable Determinations of Gestational Fetal Exposure to Drugs and Other Chemical Agent.* U.S. Patent 5,326,708. July 5, 1994.
16. Lewis DE. *Forensically Acceptable Determinations of Gestational Fetal Exposure to Drugs and Other Chemical Agents.* U.S. Patent 5,532,131. July 2, 1996.
17. Lewis DE, Moore CM. *Meconium Assay Procedure.* U.S. Patent 5,587,323. December 24, 1996.
18. Moore CM, Negrusz A. Drugs of abuse in meconium. *Forensic Sci Rev* **7**(2): 103–117; 1995.
19. Moore C, Lewis D, Negrusz A. Determination of drugs in meconium. *J Chromatogr (Biomed Applns)* **713**(1):137–146; 1998.
20. Lester BM, ElSohly M, Wright LL, Smeriglio VL, Verter J, Bauer CR, Shankaran S, Bada HS, Walls HH, Huestis MA, Finnegan LP, Maza PL. The Maternal Lifestyle Study: drug use by meconium toxicology and maternal self-report. *Pediatrics* **107**(2): 309–317; 2001.
21. Ryan RM, Wagner CL, Schultz JM, Varley J, DiPreta J, Sherer DM, Phelps DL, Kwong T. Meconium analysis for improved identification of infants exposed to cocaine in utero. *J Pediatr* **125**(3):435–440; 1994.
22. Lewis DE, Moore CM, Leikin JB, Koller A. Meconium analysis for cocaine: a validation study and comparison with paired urine analysis. *J Anal Toxicol* **19**(3):148–150; 1995.
23. Bar-Oz B, Klein J, Karaskov T, Koren G. Comparison of meconium and neonatal hair analysis for detection of gestational exposure to drugs of abuse. *Arch Dis Child Fetal Neonatal Ed* **88**:F98–F100; 2003.
24. Szeto HH. Kinetics of drug transfer to the fetus. *Clin Obstet Gynaecol* **36**: 246–254; 1993.
25. Mahone PR, Scott K, Sleggs G, D'Antoni T, Woods JR. Cocaine and metabolites in amniotic fluid may prolong fetal drug exposure. *Am J Obstet Gynecol* **171**: 465–469; 1994.

26. Schutzman DL, Frankenfield-Chernicokk M, Clatterbaugh HE, Singer J. Incidence of intrauterine cocaine exposure in a suburban setting. *Pediatrics* **88**: 825–827; 1991.

27. Ostrea EM Jr., Welch RA. Detection of prenatal drug exposure in the pregnant women and her newborn infant. *Clin Perinatol* **18**(3):629–645; 1991.

28. Chen C, Raisys V. Detection of cocaine and benzoylecgonine in meconium by enzyme multiplied immunoassay technique (EMIT) following solid-phase extraction. *Clin Chem* **38**:1008: 1992.

29. Moriya F, Chan KM, Noguchi TT, Wu PY. Testing for drugs of abuse in meconium of newborn infants. *J Anal Toxicol* **18**(1):41–45; 1994.

30. Lewis D, Moore C, Leikin JB, Kechavarz L. Multiple birth concordance of street drug assays of meconium analysis. *Vet Hum Toxicol* **37**(4):318–319; 1995.

31. Franssen RME, Stolk LML, van den Brand W, Smit BL. Analysis of morphine and amphetamine in meconium with immunoassay and HPLC-diode array detection. *J Anal Toxicol* **18**:294–295; 1994.

32. ElSohly MA, Stanford DF, Murphy TP, Lester BM, Wright LL, Smeriglio VL, Verter J, Bauer CR, Shankaran S, Bada HS, Walls HC. Immunoassay and GC-MS procedures for the analysis of drugs of abuse in meconium. *J Anal Toxicol* **23**(6):436–445; 1999.

33. Kwong TC, Ryan RM. Detection of intrauterine illicit drug exposure by newborn drug testing. *Clin Chem* **43**(1):235–242; 1997.

34. Moore CM, Lewis DE, Leikin JB. False positive and false negative results in meconium drug testing. *Clin Chem* **41**(11):1614–1616; 1995.

35. Browne S, Moore C, Negrusz A, Tebbett I, Covert R, Dusick A. Detection of cocaine, norcocaine and cocaethylene in the meconium of premature neonates. *J Forens Sci* **39**(6):1515–1519; 1994.

36. Murphey LJ, Olsen GD, Konkol RJ. Quantitation of benzoylnorecgonine and other cocaine metabolites in meconium by high-performance liquid chromatography. *J Chromatogr.* **613**(2):330–335; 1993.

37. Stolk LM, Coenradie SM, Smit BJ, van As HL. Analysis of methadone and its primary metabolite in meconium. *J Anal Toxicol* **21**(2):154–159; 1997.

38. Pichini S, Pacifici R, Pellegrini M, Marchei E, Perez-Alarcon E, Puig C, Vall O, Garcia-Algar O. Development and validation of a liquid chromatography-mass spectrometry assay for the determination of opiates and cocaine in meconium. *J Chromatogr B Analyt Technol Biomed Life Sci* **794**(2):281–292; 2003.

39. Clark GD, Rosenzweig IB, Raisys V, Callahan CM, Grant TM, Streissguth AP. The analysis of cocaine and benzoylecgonine in meconium. *J Anal Toxicol* **16**(4): 261–263; 1992.

40. Steele BW, Bandstra ES, Wu NC, Hime GW, Hearn WL. *m*-hydroxy benzoylecgonine: an important contributor to the immunoreactivity in assays for benzoylecgonine in meconium. *J Anal Toxicol* **17**:348–352; 1993.

41. Lewis D, Moore C, Becker J, Leikin J. Prevalence of meta-hydrox ybenzoylecgonine (m-OH-BZE) in meconium samples. *Bulletin of the lnt Ass Forens Toxicol* **25**(3):33–36; 1995.

42. Oyler J, Darwin WD, Preston KL, Suess P, Cone EJ. Cocaine disposition in meconium from newborns of cocaine-abusing mothers and urine of adult drug users. *J Anal Toxicol* **20**:453–462; 1996.

43. ElSohly MA, Kopycki W, Feng S, Murphey TP. Identification and analysis of cocaine metabolites in meconium. *J Anal Toxicol* **23**(6):446–451; 1999.
44. Abusada GM, Abukhalaf IK, Alford DD, Vinzon-Bautista I, Pramanik AK, Ansari NA, Manno JE, Manno BR. Solid-phase extraction and GC/MS quantitation of cocaine, ecgonine methyl ester, benzoylecgonine, and cocaethylene from meconium, whole blood, and plasma. *J Anal Toxicol* **17**(6):353–358; 1993.
45. Lewis DE, Moore CM, Leikin JB. Cocaethylene in meconium specimens. *J Toxicol Clin Toxicol* **32**(6):697–703; 1994.
46. Xia Y, Wang P, Bartlett MG, Solomon HM, Busch KL. An LC/MS/MS method for the comprehensive analysis of cocaine and cocaine metabolites in meconium. *Anal Chem* **72**(4):764–771; 2000.
47. Wang PP, Bartlett MG. Identification and quantification of cocaine N-oxide: a thermally labile metabolite of cocaine. *J Anal Toxicol* **23**(1):62–66; 1999.
48. Pichini S, Marchei E, Pacifici R, Pellegrini M, Lozano J, Garcia-Algar O. Application of a validated high performance liquid chromatography – mass spectrometry assay to analysis of meta-hydroxybenzoylecgonine and para-hydr oxybenzoylecgonine in meconium. *J Chromatogr B Analyt Technol Biomed Life Sci* **820**(1): 151–156; 2005.
49. Salem MY, Ross SA, Murphy TP, ElSohly MA. GC-MS determination of heroin metabolites in meconium: evaluation of four solid-phase extraction cartridges. *J Anal Toxicol* **25**(2):93–98; 2001.
50. Wingert WE, Feldman MS, Hee Kim M, Noble L, Hand I, Ja Yoon J. A comparison of meconium, maternal urine and neonatal urine for detection of maternal drug use during pregnancy. *J Forens Sci* **39**(1):150–158; 1994.
51. Moriya F, Chan KM, Noguchi TT, Parnassus WN. Detection of drugs-of-abuse in meconium of a stillborn baby and in stool of a deceased 41-day-old infant. *J Forensic Sci* **40**(3):505–508; 1995.
52. Moore CM, Deitermann D, Lewis DE, Leikin JB. The detection of hydrocodone in meconium: Two Case studies. *J Anal Toxicol* **19**:514–518; 1995.
53. Becker J, Moore C, Lewis D, Leikin J. Morphine detection in meconium: hydrolyzed v. non-hydrolyzed specimens. *Clin Chem* **41**(6):S114; 1995.
54. Habrdová V, Balíková M, Marešová V. *"Brown" and "Pervitin" in Newborn's Meconium: A Case Report*. The International Association of Forensic Toxicologists (TIAFT), 39th Annual International Meeting, 2001, Prague, The Czech Republic.
55. Le NL, Reiter A, Tomlinson K, Jones J, Moore C. The detection of oxycodone in meconium specimens. *J Anal Toxicol* **29**(1):54–57; 2005.
56. Rao R, Desai NS. OxyContin and neonatal abstinence syndrome. *J Perinatol* **22**:324–325; 2002.
57. ElSohly MA, Feng S, Murphy TP. Analysis of methadone and its metabolites in meconium by enzyme immunoassay (EMIT) and GC-MS. *J Anal Toxicol* **25**(1): 40–44; 2001.
58. Choo RE, Murphy CM, Jones HE, Huestis MA. Determination of meth adone, 2-ethylidene-1,5-dimethyl-3,3-diphenylpyrrolidine, 2-ethyl-5-methyl-3,3-diphenylpyraline and methadol in meconium by liquid chromatography atmospheric pressure chemical ionization tandem mass spectrometry. *J Chromatogr B Analyt Technol Biomed Life Sci* **814**(2):369–373; 2005.

59. Pichini S, Pacifici R, Pellegrini M, Marchei E, Lozano J, Murillo J, Vall O, Garcia-Algar O. Development and validation of a high-performance liquid chromatography-mass spectrometry assay for determination of amphetamine, methamphetamine, and methylenedioxy derivatives in meconium. *Anal Chem* **76**(7):2124–2132; 2004.
60. Ho E, Karimi-Tabesh L, Koren G. Characteristics of pregnant women who use ecstasy (3,4-methylenedioxymethampohetamine). *Neurotox Teratol* **23**: 561–567; 2001.
61. Tabor BL, Smith-Wallace T, Yonekura ML. Perinatal outcome associated with PCP versus cocaine use. *Am J Drug Alcohol Abuse* **16**(3–4):337–348; 1990.
62. Moore CM, Lewis DE, Leikin JB. The determination of phencyclidine in meconium. *J Forensic Sci* **41**(6):1057–1059; 1996.
63. Moore CM, Becker JW, Lewis DE, Leikin JB. The determination of 11-nor-Δ^9-tetra-hydrocannabinol-9-carboxylic acid (THC-COOH) in meconium. *J Anal Toxicol* **20**:50; 1996.
64. Feng S, ElSohly MA, Salamone S, Salem MY. Simultaneous analysis of D9-THC and its major metabolites in urine, plasma, and meconium by GC-MS using an immunoaffinity extraction procedure. *J Anal Toxicol* **24**(6):395–402; 2000.
65. ElSohly MA, Feng S. D 9-THC metabolites in meconium: identification of 11-OH-delta 9-THC, 8 beta,11-diOH-D 9-THC, and 11-nor-delta 9-THC-9-COOH as major metabolites of delta 9-THC. *J Anal Toxicol* **22**(4):329–335; 1998.
66. Goosensen M, Stolk LM, Smit BJ. Analysis of 11-nor-delta 9-THC-carboxylic acid in meconium with immunoassay and HPLC diode-array detection. *J Anal Toxicol* **19**(5):330; 1995.
67. Ostrea EM Jr, Knapp DK, Romero A, Montes M, Ostrea AR. Meconium analysis to assess fetal exposure to nicotine by active and passive maternal smoking. *J Pediatr* **124**(3):471–476; 1994.
68. Dempsey D, Moore C, Deitermann D, Lewis D, Feeley B, Niedbala RS. The detection of cotinine in hydrolyzed meconium samples. *Forensic Sci Int* **102** (2–3):167–171; 1999.
69. Baranowski J, Pochopien G, Baranowska I. Determination of nicotine, cotinine and caffeine in meconium using high-performance liquid chromatography. *J Chromatogr B Biomed Sci Appl* **707**(1–2):317–321; 1998.
70. Bearer CF, Jacobson JL, Jacobson SW, Barr D, Croxford J, Molteno CD, Viljoen DL, Marais AS, Chiodo LM, Cwik AS. Validation of a new biomarker of fetal exposure to alcohol. *J Pediatrics* **143**(4):463–469; 2003.
71. Chan D, Klein J, Koren G. Validation of meconium fatty acid ethyl esters as biomarkers for prenatal alcohol exposure. *Pediatrics* **144**(5):692; 2004.
72. Chan D, Klein J, Karaskov T, Koren G. Fetal exposure to alcohol as evidenced by fatty acid ethyl esters in meconium in the absence of maternal drinking history in pregnancy. *Ther Drug Monit* **26**(5):474–481; 2004.
73. Moore C, Jones J, Lewis D, Buchi K. Prevalence of fatty acid ethyl esters in meconium specimens. *Clin Chem* **49**(1):133–136; 2003.
74. Koren G, Chan D, Klein J, Karaskov T. Estimation of fetal exposure to drugs of abuse, environmental tobacco smoke, and ethanol. *Ther Drug Monit* **24**(1): 23–25; 2002.

75. Chan D, Caprara D, Blanchette P, Klein J, Koren G. Recent developments in meconium and hair testing methods for the confirmation of gestational exposures to alcohol and tobacco smoke. *Clin Biochem* **37**(6):429–438; 2004.
76. Derauf C, Katz AR, Easa D. Agreement between maternal self-reported ethanol intake and tobacco use during pregnancy and meconium assays for fatty acid ethyl esters and cotinine. *Am J Epidemiol* **158**(7):705–709; 2003.

Chapter 3

Drugs-of-Abuse in Nails

Diana Garside

Summary

The utility of nails in forensic toxicology for the analysis of drugs-of-abuse is examined. This chapter reviews the basic structure of the nail, mechanisms of drug incorporation, drugs-of-abuse that have been detected in nails, analytical methodologies, interpretation of results, advantages and disadvantages, and a chronological review of the literature. Drugs-of-abuse that are discussed within this chapter include amphetamine, methamphetamine, MDMA/MDA, cocaine and its metabolites, morphine, codeine, 6-acetylmorphine, hydrocodone, hydromorphone, oxycodone, methadone, cannabinoids, and phencyclidine.

Key Words: Nails, drugs-of-abuse, toxicology.

1. INTRODUCTION

Nails have long been recognized as indicators of systemic health problems by dermatologists, and the study of nails in this field is known as onychopathology. Disease states as diverse as cardiac failure and leprosy can affect the pathology of nails *(1,2)*. Nails will also be affected from the ingestion of a wide variety of exogenous agents including antibiotics, chemotherapeutic drugs, heavy metals, and antimalarial drugs *(3)*. As a matrix for detecting drug ingestion, nails are best associated with arsenic poisoning and the presence of Mees' lines *(4–8)*. Nails have also been specifically analyzed for transition metals in occupational and environmental exposure cases *(8–11)*.

From: *Forensic Science and Medicine: Drug Testing in Alternate Biological Specimens*
Edited by: A. J. Jenkins © Humana Press, Totowa, NJ

In addition, nail analysis for therapeutic drug monitoring is practiced in the clinical chemistry arena *(12–18)*. With the advent of hair testing for drugs-of-abuse, nail analysis was the natural progression for forensic toxicologists and was first reported in 1984 for the detection of methamphetamine abuse *(19)*. To date, the application of nail analysis for drugs-of-abuse has been described in antemortem and postmortem toxicology, and in prenatal drug exposure. This chapter will review the analysis of nails for drugs-of-abuse and their utility in forensic toxicology.

2. STRUCTURE OF NAILS

Before the utility of nails in forensic toxicology can be evaluated, it is necessary to understand their anatomy. Nails perform two functions: to protect and to enhance the sense of touch of the fingertip. There are six major components of the nail, namely the germinal matrix, the lunula, the nail bed, the hyponychium, the nail plate, and the nail folds *(20)* (Fig. 1).

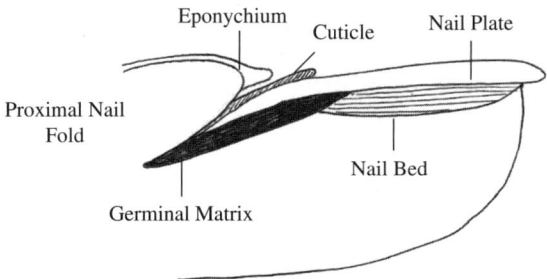

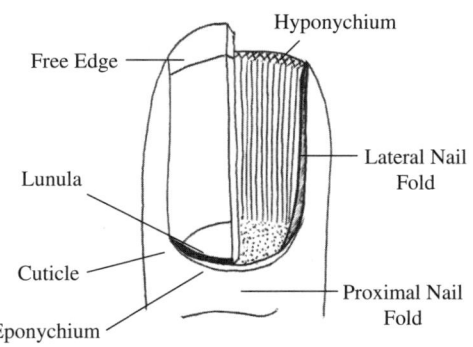

Fig. 1. Schematic of anatomy of nail.

2.1. Germinal Matrix

The germinal matrix of the nail, or the nail root, is the site of the majority of nail production. It lies beneath the skin, forms the floor of the proximal nail groove, and is protected by the proximal nail fold. The germinal matrix produces keratin cells that constitute the nail plate. As new keratin cells are formed, they push the older ones out through the cuticle where they flatten, undergo cytoplasmic condensation, and form the translucent nail plate. The germinal matrix contains few melanocytes, cells responsible for producing melanin, and hence nails lack pigmentation. (The pink color seen in the nail plate is actually the blood supply to the nail bed beneath the nail plate.) The size and shape of the germinal matrix determines the shape and thickness of the nail plate: the longer the matrix, the more cells it can produce and the thicker the nail. The shape of the matrix varies between individuals. A curved matrix will produce a curved nail plate while a flat matrix generates a flat nail.

2.2. Lunula

The lunula is the white, half-moon-shaped area seen at the base of the nail. It is most prominent in the thumbnail and is where the matrix extends beyond the proximal nail fold. It is hypothesized that the white color is a result of light reflection or of a zone of incomplete keratinization.

2.3. Nail Bed

The nail bed extends from the lunula to the hyponychium. It is the flat surface to which the nail plate adheres, is very vascular, and it consists of longitudinal epidermal ridges and dermal papillae.

2.4. Hyponychium

The area under the nail plate, where the nail bed ends and the normal epidermis begins, is referred to as the hyponychium. It is the area between the free edge of the nail plate and the fingertip, and it acts as a waterproof barrier.

2.5. Nail Plate

The nail plate is the hard, translucent material known more simply as the nail. It is composed of interlocking, dead, keratinous cells that lack nuclei and organelles. The cells are so tightly knit they form a smooth surface that is not conducive to exfoliation, unlike the skin. The nail plate is approximately 0.5 mm thick, is generally thicker in men than women, and thickens with age. The nail plate at the free edge is thinner than the plate over the nail bed *(21)*. It is the nail plate that will be discussed as an alternate biological matrix for use in forensic toxicology.

2.6. Nail Folds

The nail plate sits on the nail bed in three groves made by three folds of skin: the proximal nail fold and two lateral nail folds. The nail folds act as protective barriers against bacteria for the matrix and nail bed. The nail plate grows from under the proximal nail fold, which is an extension of the epidermis and dermis and contains sweat glands. The epithelium of the proximal nail fold, called the eponychium, attaches to the nail plate and moves with it as it grows. The cuticle is a build-up of dead skin cells that are sloughed off from the underside of the proximal nail fold as the nail plate grows out from under it and also attaches to the nail plate. Cosmetically, the eponychium and the cuticle are pushed back during a manicure.

2.7. Growth Rates

Fingernails grow at an average rate of 0.1 mm per day (3 mm per month; range 1.9–4.4 mm per month) *(20)*. It takes about 2 months for the new nail to grow the 5 mm out from under the proximal nail fold and about 6 months to completely replace a nail that has been removed. Toenails grow at about one-third the rate of fingernails. Consequently, it takes from 12 to 18 months for a new toenail to grow out. The rate of growth varies between individuals and even between digits. Generally, the rate of nail growth is proportional to the length of the digit, the "pinkie" fingernail being the slowest followed by the thumbnail. The nail on the index finger tends to have the fastest growth rate. Nails grow faster in children, their peak growth rate (up to 0.15 mm per day) occurring between the ages of 10–14 years and then slow down significantly in the second decade *(22)*. With slower growth rates, the cells in the nail plate are larger resulting in thicker nails. This correlates with an increase in thickness of the nail plate with increased age. Nails grow faster in the dominant hand and in nail biters due to external stresses and in males. Other factors such as the weather, illness, diet, and physiologic changes such as pregnancy will also determine growth rates *(20)*.

2.8. Nail Formation

It was once widely accepted that nail is formed only at the germinal matrix *(23–26)*. A second school of thought suggests that about 20% of the total nail mass is continuously being generated as the nail grows out over the nail bed *(21,27)*. Studies have shown that the nail plate increases in thickness from the proximal to distal end at a rate of 0.02 mm/mm (0.22% per each 1% increment in length) *(21)*. This latter theory would explain how a moving nail can remain attached to the nail bed and the kinetics of drug incorporation into the distal nail when the presence of an antifungal drug is detected in the distal nail plate long before it should if the drug was only incorporated during nail formation in the germinal matrix *(12,15,28,29)*.

3. Drug Incorporation

3.1. Internal Mechanisms

As there is debate as to where and precisely how the nail plate is grown *(20,21)*, the mechanism of drug incorporation in the matrix is still unclear. It is probable that there are at least two mechanisms of drug incorporation corresponding to the two mechanisms of nail formation mentioned in Section 2.8. Thus, drugs become embedded in the nail plate from the blood supplying the germinal matrix during keratinization and from the vasculature in the nail bed that also contributes to nail formation. Other routes of drug incorporation must also be considered, such as from the sweat generated in the proximal nail fold that bathes the nail plate as it passes through it during growth and longitudinal diffusion from the proximal to distal end.

Because of the difficulty of administering known amounts of drugs-of-abuse to subjects for the purposes of determining pharmacokinetics, much of the information regarding drug incorporation into nail comes from the therapeutic drug monitoring literature. There does appear to be a direct correlation between the concentration of an antifungal drug that is incorporated into nail and the dose, which also correlates with an improved cure rate *(13)*. The dose-concentration relationship is also evident with cocaine *(30)* and haloperidol *(18)*. The correlation is less evident, however, with codeine *(30)* and methadone *(31)* although the latter information relied upon self-report, a notably inaccurate measurement. In addition, incorporation of drugs appears to reach steady state with chronic use *(32)*. After a therapeutic oral dosing study of the antifungal drug, itraconazole, the drug was detected in the distal nail clippings of the toenail up to 9 months and in the fingernail clippings up to 6 months after the end of dosing *(33)*. A similar finding was observed with terbinafine *(15)*. This may exemplify the permanent incorporation of drugs into the nail matrix and their stability once present.

In paired hair and nail studies, nail contained a smaller mean peak concentration of drug for terbinafine *(34)*, haloperidol *(18)*, cocaine *(30,35)*, and codeine *(30)* per gram compared to hair although amphetamines tend to be less consistent in this observation. For example, Suzuki et al. discovered that methamphetamine was present at higher concentrations in nail than in hair approximately 50% of the time (19,36), Cirimele et al. found that amphetamines were present in approximately equal concentrations in fingernail scrapings and a paired hair sample (37), and Lin et al. demonstrated hair specimens contain higher methamphetamine and amphetamine concentrations than fingernail clippings (38). The difference in the concentration of drugs in nail compared to hair could be due to differences in the mechanism of incorporation of drugs between hair and nail, the different growth rates of the matrices, sampling the correct sections of paired nail and hair that corresponded to drug use, or differences in the structure of the matrices; nails contain no melanin, which has been shown to influence the concentration of drugs detected in hair: the more melanin present, the more drug is detected.

A study of methamphetamine disposition in nails compared to the corresponding hair segments of drug abusers undergoing treatment revealed evidence that drugs are deposited along the entire nail plate and not exclusively at the root (38). Other factors such as the physiochemical nature of the drug, its molecular weight, lipophilicity, dissociation constant, amount of protein binding, bioavailability, and route of administration are important in its ability to be incorporated into nail.

3.2. External Mechanisms

Exposure to external sources of drug such as from the environment, sweat, sebum, and saliva cannot be dismissed as viable mechanisms of drug incorporation into nail. Pounds et al. (6) suggested that the small concentration of arsenic observed in the fingernail clippings of volunteers given therapeutic amounts of arsenic approximately 1 month after ingestion was due to contamination of the ventral nail surface from sweat. This suggestion was made upon detecting the majority of the arsenic in the portion of nail that corresponded to nail formation at the root at the time of arsenic ingestion. Access through sweat was also the mechanism postulated during the oral administration of fluconazole for the treatment of onychomycosis (fungal infection of the nail) (14). The antifungal drug was detected in the nail plate after only 8 h postdose and because of the high water solubility of fluconazole, incorporation through sweat appeared to be a reasonable hypothesis. The results from one of the few controlled studies involving the disposition of drugs-of-abuse, opiates and cocaine, in hair and nails also support the hypothesis that drugs are transferred from bodily secretions, such as sweat, into these matrices (30). Incoroporation of drugs into nail from environmental contamination such as handling drugs, however, has been suggested to be minimal. In a study to evaluate this (39), cocaine analytes and opiates were analyzed in the finger and toenails, and the corresponding nail washes, of decedents known to have used these drugs. As toenails are less likely to be subjected to environmental contamination than fingernails, one would expect to see significantly less drug present in the wash of the toenails compared to that found in the washes of the fingernails. This was not observed. The wash/nail ratios of drug present were similar for both finger and toenails for all analytes except benzoylecgonine. The higher benzoylecgonine finger/toenail wash ratio was explained by the authors as possible spontaneous hydrolysis of cocaine, stuck under the nails, to benzoylecgonine. The authors concluded that drug incorporation into nail from environmental contamination is probably not a significant mechanism for three reasons: (i) toenails contain drugs despite their exposure to this type of contamination being less likely than fingernails, (ii) the similar wash/nail ratios for finger and toenails suggest that environmental contamination is minimal, and (iii) the ratio of norcocaine to cocaine in nails is higher than that observed in illicit cocaine, an observation that would not exist if cocaine had not been metabolized enzymatically to norcocaine.

Extrapolating from therapeutic drug monitoring studies, drug penetration from external sources may depend not so much on the lipophilicity or the state of ionization of the drug (although drugs were demonstrated to be absorbed to a lesser extent when ionized than unionized) as much as molecular size. Drugs with increasing molecular weight were observed to penetrate nail samples to a lesser extent than drugs with smaller molecular weights *(40)*. This may correlate with the observation that the nail plate contains intercellular channels of finite size that leads to a certain permeability, especially to polar compounds that would otherwise be unable to pass the membranes *(20)*. In addition, as the dorsal surface of the nail plate is permeable, the medium that it is exposed to blood, sweat, sebum, saliva, water, urine, and the relative proportion of drugs contained therein will determine how much and which drugs-of-abuse and their metabolites will become incorporated in the nail through this mechanism.

The extent of drug incorporation through any of the mechanisms suggested is unknown and probably varies considerably depending on the many confounding factors involved. These include the type of drug, the physical state of the individual, and environmental conditions.

4. DRUGS DETECTED

The first reported instance of nails being utilized for the purpose of detecting illicit drug abuse was in 1984 with the detection of methamphetamine and amphetamine in the nails of methamphetamine users *(19)*. This was followed 5 years later by a second report of detecting methamphetamine and amphetamine in nails *(36)*. This study also compared these findings with the hair, sweat, and saliva of habitual methamphetamine users. The authors concluded that the concentration of methamphetamine in nails predominated its metabolite, amphetamine, and was detectable for longer periods than in the saliva and sweat of the same individuals.

It was not until the mid-1990s that nail analysis for drugs-of-abuse became more prevalent. Table 1 summarizes the literature of nail analysis for drugs-of-abuse from the first instance to date (2005). In 1994, Tiess et al. *(41)* presented a report on the analysis for cocaine in the fingernail clippings and the entire nail plate of the toe in an acute cocaine intoxication victim. Tiess et al. found comparable amounts of cocaine in the fingernail clippings of both left and right hands, which were considerably greater than in the toenail (Table 1). In addition, more cocaine was detected in the root of the toenail than the rest of the nail plate. From these observations and the results of the corresponding biological fluids and hair analysis, Tiess et al. concluded that the deceased had a chronic cocaine habit and had used the drug shortly before his death. In the same year, Miller et al. extracted cocaine and its primary metabolite, benzoylecgonine, from hair, and finger and toenails *(35)*. The findings demonstrated that cocaine and benzoylecgonine were present in nails at approximately equal concentrations but at lower concentrations than the corresponding hair.

In addition to amphetamine, 3,4-methylenedioxymethamphetamine (MDMA) and 3,4-methylenedioxyamphetamine (MDA) were detected in the scrapings of a single fingernail and in the head hair of a drug abuser *(37)*. Cirimele et al. *(37)* found the concentration of each analyte to be approximately equal in both matrices, but the amount of MDMA detected was much higher than either amphetamine or MDA (Table 1). MDMA was not present in the blood or urine from this individual. From these analyses, the authors were able to conclude that the individual tested was a chronic MDMA user, a fact not evident from the more traditional blood and urine testing.

Nails have been proposed as an optimal specimen for the detection of intrauterine drug exposure. Unless authorities suspect the mother of drug abuse before the birth, meconium may not be collected in such cases. As the window of detection in blood and urine is very small, prenatal drug use may go undetected. Nail formation begins *in utero* at the beginning of the second trimester. Any drugs absorbed by the mother during pregnancy should, therefore, be incorporated in the nail matrix through systemic exposure and through being bathed in amniotic fluid. While this is also true of hair, prenatal hair may be sparse and is lost in the perinatal period. Nails grown during gestation survive for a longer postnatal period than hair. Unless the baby is breast-fed, the presence of drugs in nail is strong evidence of prenatal exposure. The possibility of prenatal cocaine use was investigated by Skopp and Potsch *(42)*, utilizing nail analysis in a suspected case of sudden infant death syndrome (SIDS). The case involved a 3-month-old baby where no anatomical or toxicological cause of death could be found using conventional methods. A slight heart abnormality

Table 1
Chronological Summary of Drugs-of-Abuse Detected in Nails

Analyte	n (subjects)	Concentration (ng/mg nail)	State	Observation	Reference
Methamphetamine	9	0.32–17.7 (fingernail clippings) 0.06–9.93 (toenail clippings)	AM	Toe > fingernails	1984 (19)
Amphetamine		0.03–0.4 (fingernail clippings) 0.03–1.13 (toenail clippings)			
Methamphetamine	20	0.4–642 (nail)	AM	—	1989 (36)
Amphetamine		0.3–23.2 (nail)			
Cocaine	1	2200 (fingernail tips right hand) 2300 (fingernail tips left hand) 6 (toenail plate) 16 (toenail root)	PM	Finger > toenail	1994 (41)
Cocaine	—	Finger and toenails	PM	Hair > nail	1994 (35)
Benzoylecgonine	—	Finger and toenails		Cocaine ≈ BE	
Amphetamine	1	12 (fingernail scraping)	AM	Nail ≈ hair	1995 (37)
MDA		9.7 (fingernail scraping)			
MDMA		60.2 (fingernail scraping)			
Cocaine	1	~ 0.3 (finger and toenail clippings)	PM	—	1997 (42)
Benzoylecgonine		ND (finger and toenail clippings)			
Opiates		ND (finger and toenail clippings)			
Cocaine	23	0.2–140.17 (toenail clippings)	PM	—	1998 (43)
Benzoylecgonine		0.3–315.44 (toenail clippings)			
Norcocaine		0.66–6.78 (toenail clippings)			
Cocaethylene		0.73–2.6 (toenail clippings)			

(Continued)

Table 1
(Continued)

Analyte	n (subjects)	Concentration (ng/mg nail)	State	Observation	Reference
Morphine		0.16–0.72 (toenail clippings)			
Codeine		1.02–3.07 (toenail clippings)			
6-Acetylmorphine		0.41–1.7 (toenail clippings)			
hydrocodone		0.62 (toenail clippings)			
Cocaine	14	<0.25 to >10 (fingernail clippings)	PM	Finger > toenail	1998
		Trace to >10 (toenail clippings)			(44)
Benzoylecgonine		Trace to >10 (fingernail clippings)			
		Trace to >10 (toenail clippings)			
Norcocaine		Trace to 6.36 (fingernail clippings)			
		Trace to 0.87 (toenail clippings)			
Norbenzoylecgonine		0.61 to 2.59 (fingernail clippings)			
		<0.5 (toenail clippings)			
EME		0.87 to >10 (fingernail clippings)			
		Trace-1.18 (toenail clippings)			
		ND (fingernail clippings)			
EEE		<0.5 (toenail clippings)			
		<0.25 (fingernail clippings)			
Cocaethylene		1.03 (toenail clippings)			
		Trace to >10 (fingernail clippings)			
AEME		Trace-1.63 (toenail clippings)			
Cannabinoids	6	0.23–2.8 (fingernail clippings)	AM	–	1999
THC	11	0.13-6.97 (fingernail clippings)			(45)
THC-COOH	2	9.82–29.67 (fingernail clippings)			
Morphine	25	0.06–4.69 (fingernail clippings)	AM	–	2000
Morphine	22	0.14–6.9 (fingernail clippings)			(46)

Analyte	n	Concentration (ng/mg)	AM/PM	Comparison	Year (Ref)
Cocaine and mtbs	8	0.25–2.7 (fingernail washes and scrapings)	AM	Hair > fingernail	2000 (30)
Codeine	8	0.12–0.31 (fingernail washes and scrapings)		Hair > fingernail	
Methadone	27	ND- 577.8 (fingernail clippings)	AM	–	2000 (31)
	27	ND- 362.5 (fingernail clippings)			
Cocaine	15	6.0–414.1 (fingernail clippings)	PM	Finger > toenail	2002 (39)
		1.2–19.9 (toenail clippings)			
Benzoylecgonine		1.9–170.3 (fingernail clippings)			
		1.4–27 (toenail clippings)			
EME		0.4–27 (fingernail clippings)			
		0.2–3.6 (toenail clippings)			
Norcocaine		0.2–32.7 (fingernail clippings)			
		0.1–0.9 (toenail clippings)			
Cocaethylene		0.5–2.9 (fingernail clippings)			
		0.1–1.9 (toenail clippings)			
Morphine		0.08–407.9 (fingernail clippings)			
		0.05–6.9 (toenail clippings)			
Codeine		0.23–6.1 (fingernail clippings)			
		0.06–8.8 (toenail clippings)			
6-Acetylmorphine		0.19–504 (fingernail clippings)			
		0.04–13.1 (toenail clippings)			
Hydromorphone		0.21–0.28 (fingernail clippings)			
		0.12–0.45 (toenail clippings)			
Oxycodone		5.05–6.58 (fingernail clippings)			
		6.35–6.88 (toenail clippings)			

(Continued)

Table 1
(Continued)

Analyte	n (subjects)	Concentration (ng/mg nail)	State	Observation	Reference
Cocaine	17	0.1–4.6 (big-toenail plate)	PM	Toenail > hair	2004
Morphine		0.41–3.05 (big-toenail plate)		Toenail > hair	(47)
6-Acetylmorphine		0.25–0.83 (big-toenail plate)		Toenail ≈ hair	
Cocaine	2	28.7–34.5 (nails)	AM	–	2004
Benzoylecgonine		7.3–17.9 (nails)			(48)
EME		2.5–6.3 (nails)			
Cocaethylene		< LOD (nails)			
AEME		0.24–0.39 (nails)			
Amphetamine	62	<0.2–5.42 (fingernail clippings)	AM	Hair > fingernail	2004
Methamphetamine		0.46–61.5 (fingernail clippings)			(38)
Phencyclidine	3	4.13–147.9 (fingernail)	PM	Finger > toenail	2005 (49)
	4	0.33–9.74 (toenail)			

AEME, anhydroecgonine methyl ester; AM, antemortem; EEE, ecgonine ethyl ester; EME, ecgonine methyl ester; LOD, limit of detection; MDA, 3,4-methylenedioxyamphetamine; MDMA, 3,4-methylenedioxymethamphetamine; ND, not detected; PM, postmortem.

in the infant was an indicator of possible maternal drug use. Nail clippings from the fingers and toes were analyzed for cocaine and opiates in the absence of hair. Parent cocaine was detected (Table 1). While the presence of drugs in the nails of a neonate may confirm intrauterine exposure, if one can exclude environmental contamination, it cannot be used on its own to determine the cause of death. This information may be useful, however, in identifying and monitoring high-risk children and in epidemiological research for SIDS.

Toward the end of the 1990s, two studies on postmortem nail analysis were published. The first involved the study of toenails as a viable matrix for forensic toxicologists in postmortem cases. In this study, by Engelhart et al. *(43)*, cocaine and cocaine metabolites were analyzed simultaneously with opiates, and the results were compared to those of more conventional biological specimens. The majority of the nail specimens that tested positive for cocaine and/or opiates had negative toxicology results in the corresponding biological fluids. These results showed that toenails could be utilized by forensic toxicologists to gather information about a deceased person and that while recent drug use may not be reflected, a past or chronic use history could be deduced. Garside et al. *(44)* published the second of these studies, which investigated the presence of cocaine and its metabolites in the finger and toenails of decedents suspected of cocaine use (Table 1). The results revealed that 82% of the subjects tested were positive for cocaine analytes utilizing nails for the analysis. Only 30% of these cases were positive for cocaine analytes when blood or urine were the specimens of choice. Anhydroecgonine methyl ester, a marker of cocaine smoking, was one of the analytes detected in nail. Finally, although the authors also tested for *m*-hydroxybenzoylecgonine during this study, none was detected.

In 1999, the first report for the detection of cannabinoids in nails was published by Lemos et al. *(45)*. Fingernail clippings from volunteers who consumed marijuana and were attending a drug clinic were analyzed. During the study, it was determined that the extensive pre-wash procedure (Table 2) of the nails removed possible external contamination since at least one of the last three methanolic washes were negative for cannabinoids. Both RIA and GC/MS were used successfully for the detection of cannabinoids (Table 1). It was discovered, however, that the resulting basic pH of the nail digest, which was ideal for the subsequent liquid/liquid extraction of Δ9-tetrahydrocannabinol (THC), was ineffective for the recovery of the carboxylic acid metabolite, 11-nor-Δ9-tetrahydrocannabinoinl-9-carboxylic acid (THC-COOH). To recover THC-COOH, the nail digest had to be made acidic before the solvent extraction of the metabolite. There was no recovery of the parent THC under the acidic conditions.

In early 2000, the same authors who reported a method for the detection of cannabinoids in nail reported their findings of the usefulness of nail as a

Table 2
Chronological Summary of Methodology and Analytical data

Analyte	Wash protocol	Extraction	Clean-up	Deriv.	Analytical method	Ions (m/z)	LOD/ LOQ	Ref.
Amp Methamp	MeOH; water	2.5 M NaOH; 80°C	L/L	TFAA	GC/MS CI	232 246	10 pg 10 pg	(19)
Amp Methamp	5 × MeOH; 5 × water	0.6 M HCl	L/L	TFAA	EIMF	118 118	>20 pg >20 pg	(36)
Cocaine	MeOH	MeOH; 55°C 5 h	–	–	GC/MS EI; GC/NPD	–	–	(41)
Cocaine BE	MeOH	Acid	–	–	ESI MSMS	122, 304 168	–	(35)
Amp MDA MDMA	Methylene chloride	1 M NaOH; 95°C	L/L	PFPA	GC/MS EI	–	–	(37)
Cocaine	2 × MeOH	MeOH; sonicate	–	PFPA	GC/MS EI	303, 182	1 ng/10 mg	(42)
Cocaine BE Norcocaine CE Morphine 6–AM Codeine HC	3 × MeOH	0.1 M phoshate buffer (pH 5); sonicate	SPE	MSTFA	GC/MS EI	303, 182, 82 240, 361, 256 240, 256, 361 82, 196, 317 236, 287, 429 399, 340, 287 371, 178, 146 371, 370, 313	0.3 ng on-column	(43)

Analyte	Extraction	Hydrolysis	SPE	Derivatization	Detection	Ions	LOD	Ref
Cocaine	MeOH	MeOH: 40°C, 16 h	SPE	BSTFA	GC/MS EI	182, 272, 303	0.25 ng/mg	(44)
BE						82, 240, 361	0.25 ng/mg	
Norcocaine						140, 240, 346	0.25 ng/mg	
NorBE						140, 298, 404	0.50 ng/mg	
EME						82, 96, 271	0.25 ng/mg	
EEE						83, 96, 285	0.50 ng/mg	
CE						82, 196, 317	0.25 ng/mg	
AEME						152, 166, 181	0.25 ng/mg	
Canabinoids	0.1% SDS; 3 × water; 3 × MeOH	1M NaOH; 95°C, 30 min	L/L	–	RIA	–	0.5 ng/mL	(45)
THC				BSTFA	GC/MS EI	371, 386	1 ng/10 mg	
THC-COOH				BSTFA	GC/MS EI	371, 488	1 ng/10 mg	
Morphine	0.5M SDS; 3 × water; 4 × MeOH	1M NaOH; 60°C, 1-2 h	TOXI. TUBE A	–	RIA HPLC	–	-0.5 ng/10 mg	(46)
Methadone	0.1% SDS; 3 × water; 3 × MeOH	1M NaOH; 90°C, 30-40 min	– SPE	–	EIA GC/MS EI	-72	0.1 ng/10 mg 0.05ng/10 mg	(31)

(Continued)

Table 2
(Continued)

Analyte	Wash protocol	Extraction	Clean-up	Deriv.	Analytical method	Ions (m/z)	LOD/ LOQ	Ref.
Cocaine Opiates	See ref. (43)	See ref. (43)	SPE L/L	MSFTA MSTFA	GC/MS EI GC/MS EI	See ref. (43)	0.1 ng on-column	(39)
Cocaine Morphine 6-AM	2 × methylene chloride	HCl (37%); 100°C, 30 min	SPE	Proprionic anhydride	GC/MS EI	82, 182, 303, 268, 341, 397 268, 327, 383	0.1 ng/mg 0.1 ng/mg 0.1 ng/mg	(47)
Cocaine and analytes	water; 5 × methylene chloride	–	–	–	GC/MS EI	–	–	(48)
Amp Methamp	MeOH	2 M NaOH; 80°C, 1 h	L/L	HFBA	GC/MS EI	240, 118, 91 254, 210, 118	> 0.2 ng/mg> 0.2 ng/mg	(38)

6-AM, 6-acetylmorphine; BSTFA, N,O-bis(trimethylsilyl)trifluoroacetamide); CI, chemical ionization; EI, electron ionization; EIMF, electron impact mass fragmentography; ESI MSMS, electrospray ionization tandem mass spectrometry; GC/MS, gas chromatography/mass spectrometry; GC/NPD, gas chromatography/nitrogen-phosphorous detection; HFBA, heptafluorobutyric anhydride; iPrOH, isopropanol; L/L, liquid/liquid extraction; MeOH, methanol; MSTFA, N-methyl-N-trimethylsilyl-trifluoroacetamide; NaOH, sodium hydroxide; PFPA, pentafluoroproprionic acid; SDS, sodium dodecyl sulfate; SPE, solid-phase extraction; TFAA, trifluoroacetic anhydride.

matrix in the detection of morphine in known heroin users (Table 1) *(46)*. The conclusions were such that nail may be utilized to detect morphine and can, therefore, be an alternate matrix to hair in detecting past heroin use. Also in 2000, with a method for cannabinoids and morphine successfully developed, Lemos et al. applied their techniques to the analysis of nail for methadone *(31)*. Both EIA and GC/MS were effective in detecting methadone in the extracts of fingernail clippings from individuals participating in a methadone maintenance program (Table 1). Nails may be an appropriate matrix, therefore, for monitoring individual compliance in methadone maintenance programs.

The first paper to describe disposition patterns of drugs in nail was reported in 2000 *(30)*. They investigated dose-response relationships of cocaine and codeine in paired hair and nail specimens in a controlled dosing study. The study enrolled eight black males, as inpatients, who were administered codeine, and cocaine orally and by subcutaneous injection, respectively. The study was designed to simulate a short, multi-drug use session typical of regular users. The design eliminated the possibility of environmental contamination. The authors reported that decontamination washes remove more drug out of nail than hair, a dose-response relationship does exist in both hair and nail for cocaine and to a lesser degree for codeine, the concentrations of codeine and cocaine were higher in hair than in nails, and drugs appear in hair and nail within 3 days of administration. The authors acknowledged that the method of nail collection (scraping the dorsal surface) may have influenced the results.

In a study to evaluate the extent of drug incorporation into nail due to external contamination, Engelhart and Jenkins *(39)* analyzed the content of finger and toenails for drugs-of-abuse and the corresponding nail washes. Nail clippings, blood, and urine were collected from decedents who had a history of drug abuse and were analyzed for cocaine analytes and opiates. It was found that with both cocaine analytes and opiates, the concentration of drug in fingernail was significantly 7–20 and 30 times, respectively, higher than in toenails (Table 1) and the concentration of analytes in the washes of both finger and toenails was less than detected in the nails themselves. Through their observations of nail/wash ratios of finger and toenails, the authors concluded that although external contamination cannot be excluded as a route of drug incorporation into nail, it is probably not a significant mechanism of drug incorporation into nail. In addition, it was observed that there was no correlation between nail and blood concentrations.

The simultaneous detection of morphine, 6-acetylmorphine, and cocaine in toenails was reported in 2004 by Cingolani et al. *(47)*. The concentrations of the drugs found in the proximal section of the big toenail were compared to those found in the corresponding proximal hair samples. Cocaine and morphine were present in toenails at approximately twice the concentration found in hair while 6-acetylmorphine was present at approximately equal amounts (Table 1). Ragoucy-Sengler et al. *(48)* were able to detect cocaine and cocaine analytes, including anhydroecgonine methyl ester, in the nails of two subjects known to smoke cocaine (Table 1).

Another recent study evaluated the disposition of drugs into nails. Lin et al. *(38)* compared the concentration of methamphetamine in corresponding segments of nail and hair, estimated by the rate of growth of each matrix. The analyses showed that if drugs are deposited into nail exclusively at the matrix, then the proportion of drug seen in the distal clippings when compared to the corresponding portion of hair would have a similar distribution pattern. Instead, after abstinence, the concentration of methamphetamine in the distal portion of the nail decreased similarly to the new growth of hair. It was suggested that drugs must be deposited along the length of the nail bed, whereas in hair they are deposited at the scalp (root). Additional observations in this study included the concentration ratio of amphetamine to methamphetamine was the same for hair and nail, and the concentration of these analytes were greater in hair than in fingernails.

Recently, the detection of phencyclidine (PCP) in nails from four homicide victims was reported by Jenkins and Engelhart *(49)*. The corresponding blood and urine samples were also positive for PCP. Of PCP in the fingernails was generally greater than in the toenails.

5. SAMPLE PREPARATION AND ANALYSES

Analysis of nails may be described as discrete processes, namely, decontamination, preparation and extraction, clean-up, and detection. Table 2 shows a summary of the methodology and analytical data that have been utilized.

5.1. Decontamination

The first step in the analysis of nails involves washing the specimen to remove potential external contamination. Many different wash protocols have been designed including rinsing with hot or cold water, soap solution, buffer solution, and organic solvents such as methanol, isopropanol or methylene chloride, and occasionally utilizing sonication. The length of time for each wash step ranges from 15 s to 30 min. Controversy exists as to whether this step can remove all external contamination, and at what point it extracts internally incorporated drug.

5.2. *Preparation and Extraction*

Following decontamination, the analytes of interest must be extracted from the nail matrix similar to hair analysis. This may be performed by solvent extraction, or the matrix may be digested with an acid, a base, or enzymatically. Typically, the nail specimens are pulverized first or cut into small pieces to aid in the extraction process. The analytes to be extracted have to be taken into consideration during this step. For example, if cocaine is the analyte of interest, a basic digestion would be ill advised as it would result in hydrolysis to benzoylecgonine.

5.3. *Clean-Up*

Once the analytes have been extracted from the matrix, the extract has to be cleaned to remove the remains of the matrix and other unwanted compounds. This is achieved by solid phase or liquid/liquid extraction. If required, derivatization of the isolated analytes also takes place for optimal chromatography.

5.4. *Detection*

A variety of detection devices have been utilized in the analysis of nails. The most common technique is gas chromatography followed by mass spectrometry. High-performance liquid chromatography (HPLC) has also been employed. HPLC coupled with an electrochemical detector was used in the final step of the analysis for morphine in nail with a basic mobile phase comprised of phosphate buffer/isopropanol/tetrahydrofuran *(46)*. HPLC is likely to be used more frequently in the future.

6. INTERPRETATION

Nail analysis has proved a useful tool for detecting drug exposure in both antemortem and postmortem specimens, especially in the absence of more conventional samples and even hair. The answers to the obvious questions of how much drug and when did exposure occur are elusive, however. The major obstacle is the relative paucity of disposition studies and lack of complete understanding of the mechanisms of drug incorporation. Of the studies that have been conducted, it is difficult to compare the findings because of the lack of standardized testing methods and sampling techniques. In addition, the quality of the results in the studies that have been conducted are hampered by unreliable self-reports of drug use, limited paired specimens, and lack of controlled dosing.

It appears that drugs are incorporated along the nail bed from the lunula to the free edge in addition to the nail matrix and that external factors such as sweat and environmental contamination must be considered. This makes interpretation of timelines seemingly impossible. It is intuitative that toenails would be less exposed to environmental contamination than fingernails although they are probably bathed in more sweat than fingernails. Fingernails, although more likely to come into contact with drugs from external sources, are washed more frequently than toenails. Growth rates will also vary, as discussed above, with age, sex, health, digit, and environment. One study demonstrated a large difference in drug disposition between individuals of the same sex and race, with controlled dosing, utilizing the same analytical method and minimizing environmental contamination *(30)*. Although there appears to be a dose-response relationship for many drugs, there does not appear to be any correlation between blood concentrations and the concentration found in nail *(39)*.

Caution must be utilized in interpreting results of nail analysis to account for the possibility of external contamination or incorporation versus systemic exposure through ingestion or parenteral use. The presence of some drug metabolites is essential for distinguishing between these two scenarios. Cocaine, benzoylecgonine, and anhydroecgonine methyl ester could all be present from environmental contamination, but norcocaine and norbenzoylecgonine are indicative of internal metabolism.

7. ADVANTAGES AND DISADVANTAGES

The most obvious advantage and probably the most utilitarian application of nail analysis is the slow growth rate and, therefore, long retrospective analysis that is possible. This is especially true of the big toenail that may potentially represent 12 months of exposure *(8)*. Other advantages are the ease of and non-invasive collection of the sample (clippings), stability of the drug once incorporated in the matrix, which makes storage easy at room temperature, and the longevity of nail as a sample. Nail is preserved for thousands of years and is often the only viable specimen remaining on skeletonized remains. Not only can this be useful in contemperory forensic cases but also in historic investigations such as studying the lifestyle of ancient people through analysis of nails in mummies. The small sample size required for nail analysis, typically 10–50 mg, is also attractive, and unlike more conventional samples, nails are difficult to adulterate.

The exact mechanism(s) of incorporation of drugs into nails are unknown, which hinders interpretation much beyond qualitative answers and even then, without metabolites, systemic (ingestion or parenteral) exposure is ambiguous. 6-Acetylmorphine can be detected in nail, however, which allows for the determination of heroin exposure compared with morphine or codeine and provides an alternate matrix if urine is unavailable to make this definitive interpretation.

The presence of cocaethylene and anhydroecgoninemethyl ester also provides additional information. The absence of more information and a proficiency testing program is a distinct disadvantage to the reliability of the nail analysis test results.

REFERENCES

1. Holtzberg M. Nail signs of systemic disease. In: Hordinsky MK, Sawaya ME, Scher KR, eds. Atlas of Hair and Nails. Philadelphia, PA: Churchill Livingstone, 2000;59–70.
2. Daniel CR III. Nail pigmentation abnormalities. *Dermatol Clin* 1985;3(3):431–443.
3. Daniel CR III, Scher RK. Nail changes caused by systemic drugs or ingestants. *Dermatol Clin* 1985;3(3):491–500.
4. Lander H, Hodge PR, Crisp CS. Arsenic in the hair and nails. Its significance in acute arsenical poisoning. *J Forensic Med* 1965;12(2):52–67.
5. Shapiro HA. Arsenic content of human hair and nails. Its interpretation. *J Forensic Med* 1967;14(2):65–71.
6. Pounds CA, Pearson EF, Turner TD. Arsenic in fingernails. *J Forensic Sci Soc* 1979(3);19:165–173.
7. Pirl JN, Townsend GF, Valaitis AK, Grohlich D, Spikes JJ. Death by arsenic: a comparative evaluationof exhumed body tissues in the presence of external contamination. *J Anal Toxicol* 1983;7:216–219.
8. Daniel CR III, Piraccini BM, Tosti A. The nail and hair in forensic science. *J Am Acad Dermatol* 2004;50(2):258–261.
9. Gerhardsson L, Englyst V, Lundstrom NG, Norberg G, Sandberg S, Steinvall F. Lead in tissues of deceased lead smelter workers. *J Trace Elem Med Biol* 1995;9(3):136–143.
10. Wilhelm M, Hafner D, Lombeck I, Ohnesorge FK. Monitoring of cadmium, copper, lead and zinc status in young children using toenails: comparison with scalp hair. *Sci Total Environ* 1991;103(2-3):199–207.
11. Alexiou D, Koutselinis A, Manolidis C, Boukis D, Papadatos J, Papadatos C. The content of trace elements (Cu, Zn, Fe, Mg) in fingernails of children. *Dermatologica* 1980;160(6):380–382.
12. Matthieu L, De Doncker P, Cauwenbergh G, Woestenborghs R, van de Velde V, Janssen PA, Dockx P. Itraconazole penetrates the nail via the nail matrix and the nail bed: an investigation in onychomycosis. *Clin Exp Dermatol* 1991;16(5): 374–376.
13. Willemsen M, De Doncker P, Willems J, Woestenborghs R, Van de Velde V, Heykants J, Van Cutsem J, Cauwenbergh G, Roseeuw D. Posttreatment itraconazole levels in the nail. New implications for treatment in onychomycosis. *J Am Acad Dermatol.* 1992;26(5 Pt 1):731–735.
14. Laufen H, Zimmermann T, Yeates RA, Schumacher T, Wildfeuer A. The uptake of fluconazole in finger and toe nails. *Int J Clin Pharmacol Ther* 1999;37(7):352–360.
15. Dykes PJ, Thomas R, Finlay AY. Determination of terbinafine in nail samples during systemic treatment for onychomycoses. *Br J Dermatol* 1990;123(4): 481–486.

16. Schafer-Korting M. Pharmacokinetic optimisation of oral antifungal therapy. *Clin Pharmacokinet* 1993;25(4):329–341.

17. Schatz F, Brautigam M, Dobrowolski E, Effendy I, Haberl H, Mensing H, Weidinger G, Stutz A. Nail incorporation kinetics of terbinafine in onychomycosis patients. *Clin Exp Dermatol* 1995 ;20(5):377–383.

18. Uematsu T, Sato R, Suzuki K, Yamaguchi S, Nakashima M. Human scalp hair as evidence of individual dosage history of haloperidol: method and retrospective study. *Eur J Clin Pharmacol* 1989;37(3):239–344.

19. Suzuki O, Hattori H, Asano M. Nails as useful materials for detection of methamphetamine or amphetamine abuse. *Forensic Sci Int* 1984;24(1):9–16.

20. Fleckman P. Anatomy and physiology of the nail. *Dermatol Clin* 1985;3(3): 373–381.

21. Johnson M, Shuster S. Continuous formation of nail along the bed. *Br J Dermatol* 1993;128(3):277–280.

22. Hamilton JB, Terada H, Mestler GE. Studies of growth throughout the lifespan in Japanese: growth and size of nails and their relationship to age, sex, heredity, and other factors. *J Gerontol* 1955;10(4):401–415.

23. Jemec GB, Serup J. Ultrasound structure of the human nail plate. *Arch Dermatol* 1989;125(5):643–646.

24. Zaias N. The movement of the nail bed. *J Invest Dermatol* 1967 ;48(4):402–403.

25. Zaias N, Alvarez J. The formation of the primate nail plate. An autoradiographic study in squirrel monkey. *J Invest Dermatol* 1968;51(2):120–136.

26. Norton LA. Incorporation of thymidine-methyl-H3 and glycine-2-H3 in the nail matrix and bed of humans. *J Invest Dermatol* 1971;56(1):61–68.

27. Johnson M, Comaish JS, Shuster S. Nail is produced by the normal nail bed: a controversy resolved. *Br J Dermatol* 1991;125(1):27–29.

28. Munro CS, Rees JL, Shuster S. The unexpectedly rapid response of fungal nail infection to short duration therapy. *Acta Derm Venereol* 1992;72(2):131–133.

29. Shuster S, Munro CS. Single dose treatment of fungal nail disease. *Lancet* 1992;339(8800):1066.

30. Ropero-Miller JD, Goldberger BA, Cone EJ, Joseph RE Jr. The disposition of cocaine and opiate analytes in hair and fingernails of humans following cocaine and codeine administration. *J Anal Toxicol* 2000;24(7):496–508.

31. Lemos NP, Anderson RA, Robertson JR. The analysis of methadone in nail clippings from patients in a methadone-maintenance program. J Anal Toxicol. 2000;24(7):656–660.

32. Palmeri A, Pichini S, Pacifici R, Zuccaro P, Lopez A. Drugs in nails: physiology, pharmacokinetics and forensic toxicology. *Clin Pharmacokinet* 2000;38(2):95–110.

33. De Doncker P, Decroix J, Pierard GE, Roelant D, Woestenborghs R, Jacqmin P, Odds F, Heremans A, Dockx P, Roseeuw D. Antifungal pulse therapy for onychomycosis. A pharmacokinetic and pharmacodynamic investigation of monthly cycles of 1-week pulse therapy with itraconazole. *Arch Dermatol* 1996;132(1):34–41.

34. Faergemann J, Zehender H, Denouel J, Millerioux L. Levels of terbinafine in plasma, stratum corneum, dermis-epidermis (without stratum corneum), sebum, hair and nails during and after 250 mg terbinafine orally once per day for four weeks. *Acta Derm Venereol* 1993;73(4):305–309.

35. Miller M, Martz R, Donnelly B. *Drugs in Keratin Samples from Hair, Fingernails and Toenails* (abstract). Second International Meeting on Clinical and Forensic Aspect of Hair Analysis; Jun 6–8, 1994, Genoa, Italy.
36. Suzuki S, Inoue T, Hori H, Inayama S. Analysis of methamphetamine in hair, nail, sweat, and saliva by mass fragmentography. *J Anal Toxicol* 1989;13(3):176–178.
37. Cirimele V, Kintz P, Mangin P. Detection of amphetamines in fingernails: an alternative to hair analysis. *Arch Toxicol*. 1995;70(1):68–69.
38. Lin D-L, Yin R-M, Liu H-C, Wang C-Y, Liu RH. Deposition characteristics of methamphetamine and amphetamine in fingernail clippings and hair sections. *J Anal Toxicol*. 2004; 28(6):411–417.
39. Engelhart DA, Jenkins AJ. Detection of cocaine analytes and opiates in nails from postmortem cases. *J Anal Toxicol* 2002;26(7):489–492.
40. Kobayashi Y, Komatsu T, Sumi M, Numajiri S, Miyamoto M, Kobayashi D, Sugibayashi K, Morimoto Y. In vitro permeation of several drugs through the human nail plate: relationship between physicochemical properties and nail permeability of drugs. *Eur J Pharm Sci* 2004;21(4):471–477.
41. Tiess D, Wegener R, Rudolph I, Steffen U, Tiefenbach B, Weirich V, Zack F. Cocaine and benzoylecgonine concentrations in hair, nails and tissues: a comperative study of ante and postmortem materials in a case of an acute lethal cocaine intoxication. In: Tampa, FL, V. Spiehler, eds. *Proceedings of the TIAFT/SOFT Joint Congress on Forensic Toxicology, 1994*. Newport Beach, CA, 1995, pp 343–344.
42. Skopp G, Potsch L. A case report on drug screening of nail clippings to detect prenatal drug exposure. *Ther Drug Monit* 1997 ;19(4):386–389.
43. Engelhart DA, Lavins ES, Sutheimer CA. Detection of drugs of abuse in nails. *J Anal Toxicol* 1998; 22(4):314–318.
44. Garside D, Ropero-Miller JD, Goldberger BA, Hamilton WF, Maples WR. Identification of cocaine analytes in fingernail and toenail specimens. *J Forensic Sci* 1998;43(5):974–979.
45. Lemos NP, Anderson RA, Robertson JR. Nail analysis for drugs of abuse:extraction and determination of cannabis in fingernails by RIA and GC-MS. *J Anal Toxicol* 1999;23(3):147–152.
46. Lemos NP, Anderson RA, Valentini R, Tagliaro F, Scott RT. Analysis of morphine by RIA and HPLC in fingernail clippings obtained from heroin users. *J Forensic Sci* 2000;45(2):407–412.
47. Cingolani M, Scavella S, Mencarelli R, Mirtella D, Froldi R, Rodriguez D. Simultaneous detection and quantitation of morphine, 6-acetylmorphine, and cocaine in toenails: comparison with hair analysis. *J Anal Toxicol* 2004;28(2):128–131.
48. Ragoucy-Sengler C, Kintz P. Detection of smoked cocaine marker (anhydroecgonine methylester) in nails. In: M. LeBeau, ed. *Proceedings of the Joint SOFT/TIAFT Meeting*, Washington, DC, 2004, pp 309.
49. Jenkins AJ, Engelhart DA. Detection of PCP in Nails of Four Homicide Victims. *Personal Communication*, 2005.

Chapter 4

Drug Testing in Hair

Pascal Kintz

Summary

Given the limitations of self-reports on drug use, testing for drugs-of-abuse is important for most clinical and forensic toxicological situations, both for assessing the reality of the intoxication and for evaluation of the level of drug impairment. It is generally accepted that chemical testing of biological fluids is the most objective means of diagnosis of drug use. The presence of a drug analyte in a biological specimen can be used to document exposure. The standard in drug testing is the immunoassay screen, followed by the gas chromatographic–mass spectrometric confirmation conducted on a urine sample. In recent years, remarkable advances in sensitive analytical techniques have enabled the analysis of drugs in unconventional biological specimens such as hair. The advantages of this sample over traditional media, like urine and blood, are obvious: collection is noninvasive, relatively easy to perform, and in forensic situations, it may be achieved under close supervision of law enforcement officers to prevent adulteration or substitution. The window of drug detection is dramatically extended to weeks, months, or even years when testing hair. It appears that the value of analysis of alternative specimens for the identification of drug users is steadily gaining recognition. This can be seen from its growing use in pre-employment screening, in the forensic sciences, in clinical applications, and for doping control. Hair analysis may be a useful adjunct to conventional drug testing in urine. Methods for evading urinalysis do not affect hair analysis. The aim of this chapter is to document toxicological applications of hair analysis in drug detection.

Key Words: Hair, addiction, drug, toxicology, analysis, history, forensic, long-term abuse.

From: *Forensic Science and Medicine: Drug Testing in Alternate Biological Specimens*
Edited by: A. J. Jenkins © Humana Press, Totowa, NJ

1. INTRODUCTION

In the 1960s and 1970s, hair analysis was initially used to evaluate exposure to heavy metals, such as arsenic, lead, or mercury. Researchers hypothesized that hair was a preferable specimen to blood and urine for evaluating environmental hazards, because hair could store substances for an extended time period. Therefore, analyses of whole or segmented hairs might provide objective data on the extent of an individual's exposure to metals and the potential for harm. Examination of hair for organic substances, especially pharmaceuticals and drugs-of-abuse, occurred several years later because of the lack of sensitivity of analytical methods for these compounds. In 1979, Baumgartner and colleagues *(1)* published the first report using radiommunoassay (RIA) to detect morphine in the hair of heroin abusers. This paper was followed by many studies that mostly utilized RIA and/or gas chromatography/mass spectrometry (GC/MS). Today, chromatographic procedures, especially those coupled to MS, represent the gold standard for the identification and quantification of drugs in hair because of their separation ability and sensitivity.

Technically, testing of hair for drugs is not more difficult or challenging than testing in many other matrices. In fact, the application of analytical methods and instrumental approaches is in most cases quite similar, regardless of the initial sample preparation. Today, hair analysis is routinely used as a tool for detection of xenobiotics (drugs-of-abuse, pharmaceuticals, environmental contaminants, hormones, etc.) in forensic science, traffic medicine, occupational medicine, and clinical toxicology. After a discussion of drug incorporation into this matrix, this view aims to summarize and discuss the various applications of hair analysis.

2. HAIR COMPOSITION

Hair, a product of differentiated organs in the skin of mammals, is composed of protein (65–95%, keratin essentially), water (15–35%), lipids (1–9%), and minerals (<1%). The hair shaft consists of an outer cuticle that surrounds a cortex, and the cortex surrounding a central medulla in some types of hair. The hair shaft begins in a follicle closely associated with glands (sebaceous and apocrine) and grows in cycles, alternating between periods of growth (*anagen* phase) and periods of quiescence (*catagen* and *telogen* phases). Of the approximately 1 million hair follicles of the adult human scalp, approximately 85% of the hair is in the growing phase and the remaining 15% is in a quiescent stage at any time. Head hair is produced for 4–8 years at a rate of approximately 0.22–0.52 mm/day or 0.6–1.42 cm/month *(2)*. Hair from other locations typically grows for 6 months. The growth rate depends on hair type (racial difference), age, gender and anatomical location. It is considered that the vertex posterior region of the head provides least variability in growth rate.

3. DRUG INCORPORATION

It is generally proposed that drugs may enter into hair by two processes: adsorption from the external environment and incorporation into the growing hair shaft from blood supplying the hair follicle. Drugs can enter the hair from exposure to chemicals in aerosols, smoke, or secretions from sweat and sebaceous glands. Sweat is known to contain drugs present in blood. Because hair is very porous and can increase its weight up to 18% by absorbing liquids, drugs may be transferred easily to hair through sweat. Finally, chemicals present in air (smoke, vapors, etc.) can be deposited onto hair.

Drugs appear to be incorporated into the hair by at least three methods: from the blood during hair formation from sweat and sebum and from external environment. This model is more able than a passive model (transfer from blood into the growing cells of the hair follicle) to explain several experimental findings such as the following: (i) drug and metabolite(s) ratios in blood are quite different than those found in hair and (ii) drug and metabolite(s) concentrations in hair differ markedly in individuals receiving the same dose. Evidence for the transfer of the drug through sweat and sebum can be suggested as drugs and metabolites are present in sweat and sebum at high concentrations and persist in these secretions longer than in blood.

The exact mechanism by which chemicals are bound into hair is not known. It has been suggested that passive diffusion may be augmented by drug binding to intracellular components of the hair cells such as the hair pigment, melanin. For example, the codeine concentration in hair after oral administration is dependent on melanin content *(3)*. However, this is probably not the only mechanism as drugs are trapped into the hair of albino animals, which lack melanin. Another mechanism proposed is the binding of drugs with sulfhydryl-containing amino acids present in hair. There is an abundance of amino acids such as cystine in hair which form cross-linking S–S bonds to stabilize the protein fiber network. Drugs diffusing into hair cells could be bound in this way.

From various studies, it has been demonstrated that after the same dose, black hair incorporates much more drugs than blond hair *(4,5)*. This has resulted in discussions about a possible racial bias in hair analysis and is still under evaluation.

After incorporation in the hair shaft, organic substances are capable of surviving for hundreds of years under favorable conditions (protected from light and humidity). For example, the cocaine metabolite, benzoylecgonine, was detected in the hair of ancient, mummified human remains of 163 individual samples. These samples were obtained from populations living in northern Chile in the past 4000 years *(6)*.

4. SPECIMEN COLLECTION AND PREPARATION

Collection procedures for hair analysis for drugs have not been standardized. In most published studies, the samples are obtained from random locations on the scalp. However, hair is best collected from the area at the back of the head, called the *vertex posterior*. Compared with other areas of the head, this area has less variability in the hair growth rate, the number of hairs in the growing phase is more constant, and the hair is less subject to age- and sex-related influences. The sample size varies considerably among laboratories and depends on the drug to be analyzed and the test methodology. Sample sizes reported in the literature range from a single hair to 200 mg. When sectional analysis is performed, the hair is cut into segments of about 1, 2, or 3 cm, which corresponds approximately to about 1, 2, or 3 months' growth. When scalp hair is not available, other types of hair (pubic hair, arm hair, or axillary hair) can be collected as an alternative source for drug detection. The collection procedure used in the author's laboratory is provided in Fig. 1.

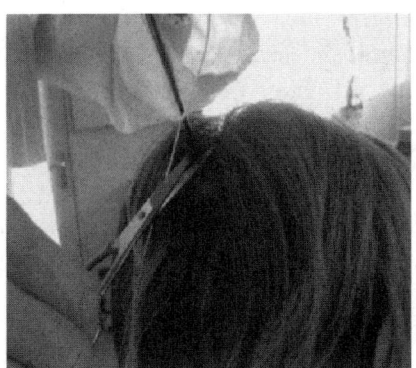 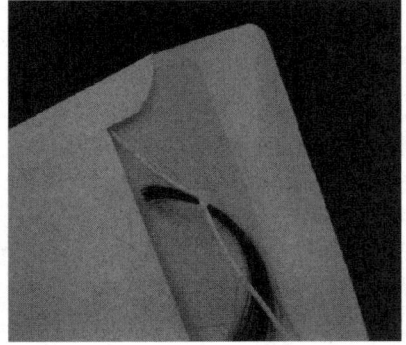

Fig. 1. Hair collection procedure (ChemTox). (i) When: 3–5 weeks after the doping control in case of challenging an urinary result or in case of drug facilitated crime upon request in other cases (drugs of abuse); (ii) how much: four strands of hair or about 100 mg hair; (iii) where: in vertex posterior; and (iv) how: root and tip ends must be distinguished, using a string 1 cm from the root and the hair must be cut by scissors as close as possible from the scalp. Note: Do not pull out! Do not use adhesive! Store in an envelope at ambient temperature.

Contaminants of hair would be a problem if they were drugs-of-abuse or their metabolites or if they interfered with the analysis and interpretation of the test results. It is unlikely that anyone would intentionally or accidentally apply anything to their hair that would contain a drug of abuse. The most crucial issue facing hair analysis is the avoidance of technical and evidentiary false-positives. Technical false-positives are caused by errors in the collection, processing, and analysis of specimens, whereas evidentiary false-positives are caused by passive exposure to the drug. Various approaches for preventing evidentiary false-positives due to external contamination of the hair specimens have been described.

The majority of laboratories use a wash step; however, there is no consensus or uniformity in the washing procedures. Among the agents used in washing are detergents such as Prell shampoo, surgical scrubbing solutions, surfactants such as 0.1% sodium dodecylsulfate, phosphate buffer, or organic solvents such as acetone, diethyl ether, methanol, ethanol, dichloromethane, hexane, or pentane of various volumes for various contact times. Generally, a single washing step is utilized, sometimes a second identical wash is performed. If external contamination is found by analyzing the wash solution, the washout kinetics of repeated washing may demonstrate that contamination has been removed.

Table 1
Recommended Limits of Quantitation (GC/MS) and Expected Concentrations for Drugs-of-Abuse in Hair

Drug	Recommended limits of quantitation	Expected concentrations
Heroin	0.2 ng/mg of 6-acetylmorphine	0.5–100 ng/mg, in most cases <15 ng/mg
Cocaine	0.5 ng/mg of cocaine	0.5–100 ng/mg, in most cases <50 ng/mg and in crack abusers >300 ng/mg is possible
Amphetamine and MDMA	0.2 ng/mg for both drugs	0.5–50.0 ng/mg
Cannabis	0.1 ng/mg for THC and 0.2 pg/mg for THC-COOH	THC, 0.05–10 ng/mg, in most cases <3 ng/mg and THC-COOH, 0.1–10 pg/mg, in most cases <1 pg/mg

From Society of Hair Testing, Recommendations for hair testing in forensic cases, *Forensic Sci Int*, 1997;84:3–6.

Table 2
Most Used Screening Procedures for the Detection of Illicit Drugs in Hair

Method	Kauert	Moeller	Kintz
Analytes	Heroin, 6-AM, dihydrocodeine, codeine, methadone, THC, cocaine, amphetamine, MDMA, MDEA, and MDA	6-AM, dihydrocodeine, codeine, methadone, THC, cocaine, amphetamine, MDMA, MDEA, and MDA	6-AM, codeine, morphine, cocaine and metabolites, amphetamine, MDMA, MDEA, MDA, and most pharmaceuticals, narcotics
Decontamination step	ultrasonic 5 min each in 5 ml H_2O, 5 ml acetone, and 5 ml petrolether	20 ml H_2O (2×) and 20 ml acetone	5 ml dichloromethane (2 × 5 min)
Homogenization	100 mg hair cut into small sections in a 30-mL vial	Ball mill	Ball mill of segments of less than 1 mm
Extraction	4 ml methanol ultrasonic 5 h, 50 °C	20–30 mg powdered hair, 2 ml acetate buffer + ß-glucuronidase/aryl-sulfatase, 90 min/40 °C	30–50 mg powdered hair and 1 ml 0.1 N HCl, 16 h/56 °C
Clean-up	None	$NaHCO_3$, SPE (C18), elution with 2 ml acetone/CH_2Cl_2 (3:1)	$(NH_4)_2HPO_4$: extract10 ml $CHCl_3$/2-propanol/n-heptane (50:17:33), organic phase purify with 0.2 N HCl, HCl phase to pH 8.4, re-extraction with $CHCl_3$
Derivatization	Propionic acid anhydride	1000 μL PFPA/75 μL PF-n-propanol, 30 min/60 °C, N_2/60 °C, 50 μL ethylacetate	40 μL BSTFA/1%TMCS or HFBA and 20 min/70 °C

Note: Today, chlorform in the procedure of Kintz has been changed to dichloromethane.
Source: From the review of Sachs and Kintz (27).

Detection of drug metabolite(s) in hair, whose presence could not be explained by hydrolysis or environmental exposure, would unequivocally establish that internal drug exposure had occurred *(7)*. Cocaethylene and nor-cocaine would appear to meet these criteria, as these metabolites are only formed when cocaine is metabolized. Because these metabolites are not found in illicit cocaine samples, they would not be present in hair as a result of environmental contamination, and thus their presence in hair could be considered a marker of cocaine exposure. This concept may be extended to other drugs, such as THC-COOH for cannabis *(8)*.

However, there is still controversy about the potential risk of external contamination, particularly for crack, cannabis, and heroin when smoked as several authors have demonstrated that it is not possible to fully deduct due to contamination *(9,10)*. In conclusion, although it is highly recommended to include a decontamination step, there is no consensus on which procedure performs best. Therefore, each laboratory must validate its own technique.

After decontamination, the hair sample can be pulverized in a ball-mill or cut into segments before the hydrolysis step or dissolved in alkaline medium (NaOH) to enhance drug solubilization. Finally, the xenobiotics are extracted or purified from the incubation medium before the analysis. The first publication describing the analysis of morphine in hair for determining opiate abuse histories reported on the analysis with RIA *(1)*. This paper was followed by a large number of studies, which mostly included RIA and/or GC/MS. Chromatographic procedures seem to be a more powerful tool for the identification and quantification of drugs in hair, owing to their separation ability and high sensitivity. Chromatographic techniques [GC or high-performance liquid chromatography (HPLC)] coupled to MS or tandem MS represent the gold standard in hair analysis for xenobiotics. The Society of Hair Testing has published recommendations for hair testing in forensic cases that include cut-off concentrations (Table 1).

Numerous papers have been published regarding the analytical aspects of drug detection. These papers include methods to test for opioids *(11–13)*, cocaine *(14–16)*, cannabis *(17–19)*, amphetamines *(20–22)*, phencyclidine *(23)*, gamma-hydroxybutyric acid (GHB) *(24)*, or benzodiazepines *(25,26)*. Details of several techniques are provided in Table 2.

5. ADVANTAGES AND DISADVANTAGES

5.1. Comparison with Urine Testing

There are essentially three problems with drug testing in urine: false-positives, instability of some compounds, and evasive maneuvers on the part of the subject such as adulteration and substitution. These problems can be reduced or eliminated using hair analysis. It is possible to obtain a fresh, identical hair sample if there is a claim of specimen mix-up or breach in the chain of custody. When head hair is missing (shaved head, bold subject), it is possible to collect alternative sources of hair, such as pubic or arm hair. This makes hair analysis essentially fail-safe in contrast to urinalysis, as an identical urine specimen cannot be obtained at a later date. To the best of the author's knowledge, no claims of external contamination have been addressed in France in any forensic situation.

Another potential use of hair analysis is to verify accidental or unintentional ingestion of liquid or food laced with drugs. In the case of single use, the hair will not test positive. Ingestion of poppy seeds appears to be sufficient for the creation of a positive opiate result in urine, whereas ingestion of up to 30 g of poppy seeds did not result in a positive hair result (Sachs, personnel communication, 1994). Its greatest use, however, may be in identifying false-negatives from urine testing, as neither abstaining from a drug for a few days nor trying to "beat the test" by diluting urine will alter the concentration in hair. Urine testing does not provide information relating to the frequency of drug use in subjects who might deliberately abstain for several days before biomedical screenings. While analysis of urine specimens cannot distinguish between chronic use or single exposure, hair analysis can make this distinction. Table 3 illustrates the differences between hair and urine as drug testing matrices.

5.2. Verification of Drug-Use History

By providing information on exposure to drugs over time, hair analysis may be useful in verifying self-reported histories of drug use in any situation

Table 3
Comparison Between Urine and Hair

Parameters	Urine	Hair
Major compound	Metabolites	Parent drug
Detection period	2–5 days	Weeks, months
Type of measure	Incremental	Cumulative
Screening	Yes	Difficult
Invasiveness	High	Low
Storage	–20 °C	Ambient temperature
Risk of false-negative	High	Low
Risk of false-positive	Low	Undetermined
Risk of adulteration	High	Low
Control material	Yes	Needed

in which a history of past rather than recent drug use is desired, such as in pre-employment and employee drug testing. Hair drug testing will detect use in an addict. In the case of an addict who takes drugs only every few days, a urine and/or blood test may be negative even when the tests are repeated.

Hair analysis can also provide a retrospective history of an individual's drug use. For this work, multi-sectional analysis is required and involves taking a length of hair and cutting it into sections to measure drug use during shorter periods of time. The hair must be cut as close as possible to the scalp and particular care is also required to ensure that the individual hairs in the cut-off tuft retain the position they originally had beside one another. For example, one can perform multi-section analysis for people who test positive on an initial screen. This information can be used to validate an individual's claim of prior drug use or abstinence during the most recent several months.

The most extensive studies on sectional analysis for drugs-of-abuse involved patients in rehabilitation centers. Segmental hair analysis was used to verify both previous drug history and recent enforced abstinence. In the instance of drug replacement therapy, the lowest drug of abuse concentrations were measured in the segments nearest the root, thus confirming decreased drug use over time or recent abstinence.

As illicit heroin samples contain codeine or codeine derivatives, codeine may also be detected in cases of heroin abuse. Morphine is a metabolite of codeine and can be detected when codeine is abused. The differentiation of heroin users from individuals exposed to other sources of morphine alkaloids can be achieved by identifying heroin, 6-acetylmorphine *(11,28)*, or acetyl-codeine *(29)*.

5.3. Determination of Gestational Drug Exposure

The increasing number of people consuming drugs result in an increase in pregnant women under the effect of these drugs. Because of the immediate and long-term health problems, newborns exposed to drugs during pregnancy should be identified soon after birth so that appropriate intervention and follow-up can be performed. Current methods to verify drug abuse include maternal self-reported drug history (self-reports are generally unreliable), maternal urinalysis (risk of false-negative results due to the short elimination half-life of the drugs and positive results only reflects exposure during the preceding 1–3 days), and analysis of amniotic fluid, urine, or meconium of the baby at time of delivery (qualitative test at the moment of delivery, risk of false-negative results due to abstinence during the preceding 1 or 3 days or limitations of testing technology).

Testing the hair of a newborn allows an increase in the window of detection (from weeks to months) and may provide information concerning the degree and pattern of the mother's drug use. Maternal drug abuse is a health hazard for the fetus, and the effects of cocaine, phencyclidine, nicotine, and other compounds are well documented. In 1987, Parton and colleagues *(30)* first reported quantitation of fetal cocaine exposure by RIA of hair obtained from 15 babies. Graham and colleagues *(31)* detected benzoylecgonine (range 0.2 to 27.5 ng/mg) in neonatal hair from seven infants whose mothers were known cocaine users. Other studies have demonstrated placental transfer of maternal drug in neonatal hair *(32)*.

5.4. Alcohol Abuse

Considering the large number of alcohol-associated problems, the diagnosis of excessive alcohol consumption is an important task from a medical point of view. The methods used for this purpose are based on indirect alcohol markers such as increased liver enzyme activity, increased erythrocyte mean cell volume, or presence of carbohydrate deficient transferrins. These markers may also originate from pathological conditions. Markers of ethanol consumption are ethyl glucuronide, and phosphatidylethanol or fatty acid ethyl esters (FAEE). The first investigation of a marker of alcohol consumption in hair was reported by Sachs and colleagues and focused on ethyl glucuronide *(33)*; however, recent examination of the presence of this ethanol metabolite in hair was rather discouraging *(34)*. Detection of ethyl glucuronide in hair is always associated with alcohol consumption, whereas a negative result does not exclude alcohol abuse. FAEE were used by Yeggles and colleagues *(35)* to monitor alcohol consumption. FAEE are formed in the presence of ethanol and free fatty acids, triglycerides, lipoproteins, or phospholipids, by a FAEE synthase found in the liver but also in hair roots. FAEE determination is of interest as they appear responsible for alcohol-induced organ damage. In blood, FAEE can be used as markers of recent alcohol intake for at least 24 h after cessation of drinking. Hair concentrations of four FAEE (ethyl myristate, ethyl palmitate, ethyl oleate, and ethyl stearate) found in hair of children, adult teetotalers, and social drinkers compared with FAEE concentrations found in hair of alcoholics led the authors to conclude that FAEE are suitable markers for the detection of heavy alcohol consumption. Segmental hair analysis in a case of alcohol withdrawal treatment showed a decrease in FAEE content from the distal to the proximal root segment *(36)*. Further studies are in progress to examine the applicability of FAEE determination in clinical practice.

5.5. Verification of Doping Practices

Athletes use both endogenous (testosterone and DHEA) and exogenous anabolic steroids (nandrolone, stanozolol, and mesterolone) because it has been claimed that they increase lean body mass, strength, and aggressiveness and lead to a shorter recovery time between workouts. Hair testing may be used to identify false-negative results due to recent abstinence for a drug a few days before a competition. Hair can also indicate the history and frequency of drug intake as repetitive abuse can be demonstrated by segmental analysis along the hair shaft in contrast to urinalysis.

A search of the international literature demonstrated a lack of articles concerning the identification of anabolic steroids in human hair *(37)*. As a complement to testosterone determination, the identification of unique testosterone esters in hair supports a charge of doping because the esters are exogenous substances *(38)*. In 2000, Thieme and colleagues *(39)* published a complete analytical strategy for detecting anabolics in hair. After extensive clean-up procedures, drugs were identified either by GLC/MS/MS or GC with high-resolution mass spectrometry (HRMS). Metandienone, stanozolol, mesterolone, metenolone enantate, nandrolone decanoate, and several testosterone esters were identified in the hair of several bodybuilders.

Another advantage of hair analysis is the possible discrimination between nandrolone and other 19 norsteroids (norandrostenedione and norandrostenediol) abuse that lead to the same urinary metabolites (norandrosterone and noretiocholanolone). Discrimination is not possible in urine, but hair can identify the parent compound. The analysis of a hair strand obtained from an athlete who tested positive for norandrosterone in urine revealed the presence of 19-norandrostenedione, distinguishing the result from nandrolone doping *(40)*. Corticosteroids *(41)* or ß-adrenergic drugs *(42)* can also be identified in hair.

5.6. Driving License Regranting

The major practical advantage of hair testing compared with urine and blood testing for drugs is the larger detection window, weeks to months, depending on the length of the hair shaft, compared with a few days for urine. Italy and Germany are using this approach in cases of driving license regranting *(45,46)*. Individuals, whose driving license has been refused or suspended for addiction to psychoactive drugs or for driving "under the influence," may claim they have stopped using drugs. They can obtain a license after a medical committee has confirmed abstinence from illicit drugs and excluded any additional risk of relapse. To provide objective evidence of abstinence from drugs with an acceptable chronological window to support the clinical decision of this medical committee, hair analysis has been included in a panel of clinical and laboratory tests aimed at retrospectively investigating the toxicological behavior of subjects. Comparison between hair analysis and urinalysis demonstrates higher diagnostic sensitivity for hair tests.

5.7. Drug-Facilitated Crimes

The use of a drug to modify a person's behavior for criminal gain is not a recent phenomenon. However, the recent increase in reports of drug-facilitated crimes (sexual assault and robbery) has caused alarm in the general public. Drugs involved can be pharmaceuticals, such as benzodiazepines, hypnotics, sedatives or anesthetics; drugs-of-abuse, such as cannabis, ecstasy, or LSD or more often ethanol. Because of low dosing, except for GHB, surreptitious administration into beverages such as coffee, soft drinks, or alcoholic cocktails is relatively simple.

Most of these substances possess amnesic properties and therefore, the victims are less able to accurately recall the circumstances under which the sexual offence occurred. As the drugs are generally short-acting, they impair an individual rapidly. In these situations, blood or even urine may not be helpful. It has been suggested that hair be tested since delays in reporting result in the removal of drug from the routine testing specimens due to metabolism. Although there are many published reports on the identification of drugs (mainly drugs-of-abuse) in hair following chronic use, those addressing controlled single dose are scarce *(45)*. The ability to differentiate between a single exposure and long-term use can be documented by multi-sectional analysis. If it is assumed that drug does not migrate along the hair shaft, a single point of exposure must be present in the segment corresponding to the period of the alleged event, using a growth rate for scalp hair of 1 cm/month. As this growth rate may vary from 0.7 to 1.4 cm/month, the length of the hair section must be calculated accordingly. A delay of 4–5 weeks between the offense and hair collection for sectional analysis (2-cm segments) can be considered as satisfactory to ensure that the hair shaft includes the time of the alleged exposure. The hair must be cut as close as possible to the scalp. Particular care is also required to ensure that the individual's hair in the strand retains the position it originally had beside another.

After decontamination, hair is then segmented as follows: 0–2 (segment corresponding to the period of alleged crime), 2–4, and 4–6 cm (which should be drug-free). Drug exposure was demonstrated by hair segmentation in several cases *(46–48)*. It was observed that the target concentrations in hair after a single exposure are generally in the range of few pg/mg. To obtain the low limits of detection required for such testing mandates the use of highly sensitive instrumentation such as tandem MS.

6. CONCLUSION

It appears that the value of hair analysis for the identification of drug users is steadily gaining recognition. This can be seen from its growing use in pre-employment screening, in forensic sciences, and in clinical applications. Hair analysis may be a useful adjunct to conventional drug testing in toxicology. Specimens can be more easily obtained with less embarrassment, and hair can provide a more long-term history of drug use.

Although there remains controversy on how to interpret the results, particularly concerning external contamination, cosmetic treatments, ethnic bias, or drug incorporation, analytical work in hair analysis has reached a plateau, with almost all the analytical problems solved. Although GLC/MS is the method of choice, in practice, GC– or LC–MS/MS technology is used today in many laboratories, even for routine cases, particularly to target low-dose compounds.

REFERENCES

1. Baumgartner AM, Jones PF, Baumgartner WA, Blank CT. Radioimmunoassay of hair for determinating opiate-abuse histories. *J Nucl Med*. 1979;**20**:748–752.
2. Saitoh M, Uzaka M, Sakamoto M, Kobori T. Rate of hair growth. In: Montana and Dobson, ed. *Advances in Biology of Skin: Hair Growth*. Oxford: 1969:183–194.
3. Kronstrand R, Förstberg-Peterson S, Kagedal B, et al. Codeine concentration in hair after oral administration is dependent on melanin content. *Clin Chem*. 1999;**45**:1485–1494.
4. Cone EJ. Mechanisms of drug incorporation in hair. *Ther Drug Monit*. 1995;**18**:438–443.
5. Henderson GL, Harkey MR, Zhou C. Incorporation of isotopically labeled cocaine into human hair: race as a factor. *J Anal Toxicol*. 1998;**22**:156–165.
6. Cartmell LW, Aufdemide AC, Spinfield A, Weems C, Arriaza B. The frequency and antiquity of prehistoric coca-leaf-chewing practices in Northen Chile: radioimmunoassay of a cocaine metabolite in hair. *Latin Am Antiquity*. 1991;**2**:260–268.
7. Cone EJ, Yousefnejad D, Darwin WD, et al. Testing human hair for drugs of abuse. I. Identification of unique cocaine metabolites in hair of drug abusers and evaluation of decontamination procedures. *J Anal Toxicol*. 1991;**15**:250–255.
8. Uhl M, Sachs H. Cannabinoids in hair: strategy to prove marijuana/hashish consumption. *Forensic Sci Int*. 2004;**145**:143–147.
9. Blank DL, Kidwell DA. Decontamination procedures for drugs of abuse in hair: are they sufficient? *Forensic Sci Int*. 1995;**70**:13–38.
10. Kidwell DA, Blank DL. In: Kintz P, ed. *Drug Testing in Hair*. Boca Raton: CRC Press, 1996:17–68.
11. Goldberger BA, Caplan YH, Maguire T, Cone EJ. Testing human hair for drugs of abuse. III. Identification of heroin and 6-acetylmorphine as indicators of heroin abuse. *J Anal Toxicol*. 1991;**15**:226–231.
12. Kintz P, Villain M, Dumestre V, Cirimele V. Evidence of addiction by anesthesiologists as documented by hair analysis. *Forensic Sci Int*. 2005;**153**:81–84.
13. Musshoff F, Lachenmeier K, Wollersen H, Lichtermann D, Madea B. Opiate concentrations in hair from subjects in a controlled heroin maintenance program and from opiate-associated fatalities. *J Anal Toxicol*. 2005;**29**:345–352.
14. Kintz P, Cirimele V, Sengler C, Mangin P. Testing human hair and urine for anhydroecgonine methylester, a pyrolysis product of cocaine. *J Anal Toxicol*. 1995;**19**:479–482.

15. Cognard E, Rudaz S, Bouchonnet S, Staub C. Analysis of cocaine and three of its metabolites in hair by gas chromatography-mass spectrometry using ion-trap detection for CI/MS/MS. *J Chromatogr B*. 2005;**826**:17–25.

16. Montagna M, Polettini A, Stramesi C, Groppi A, Vignali C. Hair analysis for opiates, cocaine and metabolites. Evaluation of a method by interlaboratory comparison. *Forensic Sci Int*. 2002;**128**:79–83.

17. Kauert G, Röhrich J. Concentrations of Δ9-tetrahydrocannabinol, cocaine and 6-acetylmorphine in hair of drug abusers. *Int J Legal Med*. 1996;**108**:294–299.

18. Cirimele V, Sachs H, Kintz P, Mangin P. Testing human hair for cannabis. III. Rapid screening procedure for the simultaneous identification of THC, cannabinol and cannabidiol. *J Anal Toxicol*. 1996;**20**:13–16.

19. Kintz P, Cirimele V, Mangin P. Testing human hair for cannabis. II. Identification of THC-COOH by GC/MS/NCI as an unique proof. *J Forensic Sci*. 1995;**40**: 619–623.

20. Kintz P, Cirimele V, Tracqui A, Mangin P. Simultaneous determination of amphetamine, methamphetamine, MDA and MDMA in human hair by GC/MS. *J Chromatogr B*. 1995;**670**:162–166.

21. Kikura R, Nakahara Y, Mieczkowski T, et al. Hair analysis for drug abuse. XV. Disposition of MDMA and its related compounds into rat hair and application to hair analysis for MDMA abuse. *Forensic Sci Int*. 1997;**84**:165–177.

22. Rothe M, Pragst F, Spiegel C, et al. Hair concentrations and self-reported abuse history of 20 amphetamine and ecstasy users. *Forensic Sci Int*. 1997;**89**:111–128.

23. Sramek JJ, W.A., Baumgartner WA, Tallos JA, Ahrens TN, Heiser JF, Blahd WH. Hair analysis for the detection of phencyclidine in newly admitted psychiatric patients. *Am J Psychiatry*. 1985;**142**:950–953.

24. Kintz P, Cirimele V, Jamey C, Ludes B. Testing for GHB in hair by GC/MS/MS after a single exposure. Application to document sexual assault. *J Forensic Sci*. 2003;**48**:195–200.

25. Villain M, Concheiro M, Cirimele V, Kintz P. Screening method for benzodiazepines and hypnotics in hair at pg/mg level by LC-MS/MS. *J Chromatogr B*. 2005;**825**:72–78.

26. Kronstrand R, Nyström I, Josefsson M, Hodgins S. Segmental ionspray LC-MS/MS analysis of benzodiazepines in hair of psychiatric patients. *J Anal Toxicol*. 2002;**26**:479–484.

27. Sachs H, Kintz P. Testing for drugs in hair. Critical review of chromatographic procedures since 1992. *J Chromatogr B*. 1998;**713**:147–161.

28. Nakahara Y, Ochiai T, Kikura R. Hair analysis for drugs of abuse. *Arch Toxicol*. 1992;**66**:446–449.

29. Kintz P, Jamey C, Cirimele V, Brenneisen R, Ludes B. Evaluation of acetylcodeine as a specific marker of illicit heroin in human hair. *J Anal Toxicol*. 1998;**22**: 425–429.

30. Parton L, Warburton D, Hill V, Baumgartner W. Quantification of fetal cocaine exposure by radio immunoassay of hair. *Pediatr Res*. 1987;**21**:372.

31. Graham K, Koren G, Klein J, Schneiderman J, Greenwald M. Determination of gestational cocaine exposure by hair analysis. *J Am Med Assoc*. 1989;**262**: 3328–3330.

32. Vinner E, Vignau J, Thibault D, Codaccioni X, Brassart C, Humbert L, Lhermitte M. Neonatal hair analysis contribution to establish a gestational drug exposure profile and predicting a withdrawal syndrome. *Ther Drug Monit.* 2003;**25**:421–432.
33. Sachs H. Drogennachweis in Haaren. In: Kijewski H, ed. *Proceedings of the Symposium on Das Haar als Spur-Spur de Haare.* Lübeck: Schmidt-Römhild, 1997:119–133.
34. Jurado C, Soriano T, Giménez MP, Menéndez M. Diagnosis of chronic alcohol consupmption. Hair analysis for ethyl-glucuronide. *Forensic Sci Int.* 2004;**145**: 161–166.
35. Yegles M, Labarthe A, Auwärter V, Hartwig S, Vater H, Wennig R, Pragst R. Comparison of ethyl-glucuronide and fatty acid ethyl ester concentrations in hair of alcoholics, social drinkers and teetotallers. *Forensic Sci Int* 2004;**145**:167–173.
36. Pragst F, Auwaerter V, Sporkert F, Spiegel K. Analysis of fatty acid ethyl esters in hair as possible markers of chronically elevated alcohol consumption by headspace solid-phase microextraction (HS-SPME) and gas chromatography-mass spectrometry (GC-MS). *Forensic Sci Int.* 2001;**121**:76–88.
37. Kintz P, Cirimele V, Ludes B. Pharmacological criteria that can affect the detection of doping agents in hair. *Forensic Sci Int.* 2000;**107**:325–334.
38. Kintz P, Cirimele V, Jeanneau T, Ludes B. Identification of testosterone and testosterone esters in human hair. *J Anal Toxicol.* 1999;**23**:352–356.
39. Thieme D, Grosse J, Sachs H, Mueller RK. Analytical strategy for detecting doping agents in hair. *Forensic Sci Int.* 2000;**107**:335–345.
40. Kintz P, Cirimele V, Ludes B. Discrimination of the nature of doping with 19-norsteroids through hair analysis. *Clin Chem.* 2000;**46**:2020–2022.
41. Cirimele V, Kintz P, Dumestre V, Goullé JP, Ludes B. Identification of ten corti-costeroids in human hair by liquid chromatography-ionspray mass spectrometry. *Forensic Sci Int.* 2000;**107**:381–388.
42. Kintz P, Dumestre-Toulet V, Jamey C, Cirimele V, Ludes B. Doping control for beta-adrenergic compounds through hair analysis. *J Forensic Sci.* 2000;**45**: 170–174.
43. Sachs H. Hair analysis as a basic for driving ability examination. *Toxicorama.* 1996;**6**:11–16.
44. Tagliaro F, De Battisti Z, Lubli G, Neri C, Manetto G, Marigo M. Integrated use of hair analysis to investigate the physical fitness to obtain the driving licence: a case-mork study. *Forensic Sci Int.* 1997;**84**:129–135.
45. Kintz P, Villain M, Cirimele V, Ludes B. Testing for the undetectable in drug-facilitated sexual assault using hair – analyzed by tandem mass spectrometry – as an evidence. *Ther Drug Monit.* 2004;**26**:211–214.
46. Villain M, Chèze M, Tracqui A, Ludes B, Kintz P. Windows of detection of zolpidem in urine and hair. Application to two drug-facilitated sexual assaults. *Forensic Sci Int.* 2004;**143**:157–161.
47. Villain M, Chèze M, Dumestre V, Ludes B, Kintz P. Hair to document drug-facilitated crimes. About 4 cases involving bromazepam. *J Anal Toxicol.* 2004;**28**:516–519.
48. Kintz P, Villain M, Chèze M, Pépin G. Identification of alprazolam in two cases of drug-facilitated incident. *Forensic Sci Int* 2005;**153**:222–226.

Chapter 5

Drugs-of-Abuse Testing in Saliva or Oral Fluid

Vina Spiehler and Gail Cooper

Summary

Historically, saliva or oral fluid has been tested for monitoring therapeutic drugs. Drugs-of-abuse, including cannabinoids, opioids, amphetamines, and cocaine, have been detected in this matrix. Multiple collection procedures and devices are available. The route of administration, collection procedures, and saliva : plasma ratios affect the amount of drug deposited. However, current analytical techniques may be utilized for testing provided the relevant compound is targeted. The window of detection is comparable to blood and is therefore conducive to detecting recent drug use.

Key Words: Saliva, drugs-of-abuse, alternate matrices.

1. INTRODUCTION

Saliva is the term used to describe the liquid excreted from more than 14 saliva glands of the head and mouth. When mixed saliva from the various glands is collected by expectoration or by placing absorbent collectors in the mouth, the resulting specimen is termed "oral fluid" and may contain blood, lymph, crevicular fluid, and cells from the gums and cheeks in addition to mixed saliva.

The use of oral fluid as a diagnostic tool has increased in recent years due to significant improvements in instrumental sensitivity and through a greater understanding of the mechanisms of drug transfer into the oral cavity. Oral fluid

From: *Forensic Science and Medicine: Drug Testing in Alternate Biological Specimens*
Edited by: A. J. Jenkins © Humana Press, Totowa, NJ

is recognized as a robust alternative testing matrix with numerous applications including monitoring for infectious diseases, monitoring compliance in drug treatment programs, and employee testing.

1.1. Historical Overview

Saliva or oral fluid has long been of interest as a non-invasive specimen that would mirror blood levels of drugs and hormones. Saliva was collected as a specimen matrix for therapeutic drug monitoring of antiepileptic drugs in children to avoid repeated blood draws in these young patients. It is a preferred specimen for studies that require repeated specimens over an extended period, such as pharmacokinetic studies, to minimize trauma to the patient *(1)*. However, the drugs of interest in monitoring are mostly acidic (antiepileptic drugs) and/or strongly protein bound, or have very low saliva/plasma ratios. Therefore, resultant saliva concentrations were below the limit of detection of analytical methods available at the time.

In 1984, Peel et al. of the Royal Canadian Mounted Police published a study of oral fluid testing for drugs involved in driving under the influence offenses *(2)*. As many illegal drugs have saliva/plasma ratios greater than 1, the oral fluid concentrations are greater than those in blood and provide the advantage of enhanced concentrations as well as the potential for correlation with impaired driving. Saliva is recognized as an advantageous specimen for roadside DUI testing *(3–7)* and criminal justice testing for drugs-of-abuse *(8–11)*.

2. COMPOSITION OF SALIVA

The largest glands and main source of saliva are the parotid, submandibular (submaximal), and sublingual glands. The parotid gland produces approximately 25% of resting saliva and 50% of stimulated saliva and produces only serous saliva. The submandibular gland (70% of resting saliva) secretes both serous and mucus saliva. The sublingual, labial, and palatal glands secrete 70% of the mucins.

Saliva is composed of 99% water, 0.7% protein (largely amylase), and 0.26% mucins. Human saliva contains two mucins: oligomeric mucous glycoprotein MG1 and monomeric mucous glycoprotein MG2. As was first recorded by Pavlov, the mucin composition of saliva depends on the nature of the stimulus for saliva secretion and which glands respond. Mechanical stimuli produce serous saliva; food stimuli produce thicker, mucous-containing saliva. In general, parasympathetic stimulation of the parotid gland leads to high rates of fluid output while sympathetic stimulation of the sublingual gland leads to high levels of protein secretion.

Salivary glands are composed of two regions: the acinar region that contains the water-permeable cells that are capable of secretion and the ductal region lined with water-impermeable cells that carry the secretions to the outlets in the mouth. Sodium and exocrine proteins are secreted by the secretory cells in the acinar region, and water passes into the lumen of the gland due to the osmotic pressure. The fluid that collects in the acinar lumen is isotonic with plasma. As the fluid travels down the saliva ducts, sodium and chloride are reabsorbed while potassium, lithium, and bicarbonate are secreted. When the fluid reaches the mouth, it is hypotonic to plasma. Resting saliva has greater acidity than plasma. When saliva moves rapidly through the ducts, less time is available for sodium reabsorption and carbonate secretion, therefore the pH of the saliva is higher *(12)*. After secretion, saliva becomes more alkaline as the dissolved carbon dioxide is lost. Saliva pH ranges from 5.6 to 7.9. In healthy donors, gingival crevicular fluid from the tooth/gum margin may constitute up to 0.5% of oral fluid collected and is of similar composition to that of plasma.

3. SAMPLE COLLECTION

Saliva can be collected using small flexible plastic cups that are applied over the opening of the saliva gland into the mouth. Collection in this manner can demonstrate the differences, if any, between drugs in saliva and drugs in oral fluid.

By definition, oral fluid is collected by absorbents placed into the mouth, by draining saliva from the mouth, or by expectoration (spitting) *(13)*. Oral fluid collected by draining or spitting is collected in a cup, vial, or test tube. A number of commercial devices employing absorbent materials are available for oral fluid collection. Absorbent collectors may be a dental cotton roll (Salivette) or plastic material placed in the mouth (Finger Collector) that is then removed and centrifuged or compressed to remove the oral fluid. More acceptable to donors are collectors consisting of an absorbent pad attached to a plastic straw or rod that the donor can place in their own mouth and then remove themselves to hand to the collector (Epitope, OraSure, Cozart RapiScan, Varian On-Site). The pad is usually placed in a preservative buffer after collection and oral fluids and/or drugs eluted by mixing with the buffer. Some devices have used an absorbent material to wipe the tongue to collect oral fluid and cells (Drugwipe).

3.1. Kinetics of Drug Transfer to Saliva/Oral Fluid

Drugs such as digoxin, steroids, and antibiotics (penicillin and tetracycline) are actively secreted into saliva. Most drugs enter saliva by passive diffusion across the cell membranes. Saliva concentrations of drugs that cross the cells by passive diffusion are related to blood or plasma concentrations of the unbound, unionized parent drug or its lipophilic metabolites *(14)*. The theoretical saliva plasma ratio (*S/P* ratio) can be calculated from the following equation derived from the Henderson–Hasselbach equation:

For basic drugs:

$$S/P = \frac{[1 + 10\,(\mathrm{pKb} - \mathrm{pHs})]}{[1 + 10\,(\mathrm{pKb} - \mathrm{pHp})]} + \frac{\mathrm{fp}}{\mathrm{fs}}$$

For Acidic Drugs:

$$S/P = \frac{[1 + 10\,(\mathrm{pHs} - \mathrm{pKa})]}{[1 + 10\,(\mathrm{pHp} - \mathrm{pKa})]} + \frac{\mathrm{fp}}{\mathrm{fs}}$$

S/P is the saliva to plasma ratio: *S* is the drug concentration in saliva and *P* is the drug concentration in plasma, pKb is the log of the ionization constant for basic drugs, pKa is the log ionization constant for acidic drugs, pHs is the pH of saliva, and pHp is the plasma pH. fp is the fraction of drug protein bound in plasma and fs is the fraction protein bound in saliva.

Drugs have been classified according to their *S/P* ratio *(14)*. Acidic drugs (pKa less than 5.5) and/or those that are highly protein bound generally have a *S/P* ratio of less than 1.0. Examples are benzodiazepines, barbiturates, and cannabinoids. Neutral drugs (pKa less than 8.5 but greater than 5.5) and alcohol have an *S/P* ratio of approximately 1.0 that does not vary with rate of saliva formation or conditions of collection. The third group includes drugs with a *S/P* ratio greater than 1. These drugs are found in higher concentrations in oral fluid than in plasma. This group includes drugs that are actively transported into oral fluid such as digoxin and penicillin as well as basic drugs (pKa greater than 8.5) and drugs with low protein binding that are ionized at saliva pH and ion trapped in the oral fluid.

Drugs-of-abuse can be classified into those that enter oral fluid by passive diffusion and those that enter oral fluid from depots in mouth tissues. In general, drugs that enter the oral fluid from depots in the oral tissues are found in higher concentrations than expected from their theoretical *S/P* ratios calculated from the Henderson–Hasselbach equation. Drugs that are abused are often smoked, creating substantial oral tissue depots. *S/P* ratios elevated more than a 100-fold after smoking have been reported for cocaine *(15)*, heroin *(15,16)*, cannabis *(17–21)*, methamphetamine *(22)* and phencyclidine (PCP) *(23,24)*. Drugs that are administered by sublingual absorption, such as fentanyl or buprenorphine, and those given as liquid preparations (codeine, morphine, and methadone) would be expected to form oral tissue depots and therefore have elevated *S/P* ratios after oral dosing *(25)*. Drugs normally taken in tablet form can also have elevated *S/P* ratios if the pill material is held in the mouth or if pill fragments remain in the mouth. This has been reported for "ecstasy" *(26,27)* and for phenytoin *(28)*. Insufflated (snorted) drugs, for example, cocaine *(10)* and heroin *(18)*, also form tissue depots bathed by oral fluid.

3.2. Effect of Collection, Collectors, and Stimulation on Drug Content of Saliva/Oral Fluid

Oral fluid has been collected from donors by expectoration, draining, absorption, and suction. During collection, saliva production can be stimulated by citrate salts or acidic sour candy or by mechanical means such as chewing paraffin or rubber bands. The presence of an absorptive collector in the mouth will stimulate saliva flow. The stimulation method will influence the mucin content and pH of the saliva. By stimulating saliva flow and increasing the saliva pH, the drug content of oral fluid is expected to be reduced for basic drugs as per the Henderson–Hasselbach equation *(29)*. Collection devices can also affect recovery of drugs from oral fluid due to retention of oral fluid in the absorbent, adsorption of the drug on the device components, and the effectiveness of device buffers to release drugs from the collector *(25,30)*. After controlled administration of codeine, oral fluid codeine concentrations were higher in specimens obtained by expectoration or draining of unstimulated oral fluid than those obtained by absorptive devices *(25)* and were higher in specimens collected without stimulation than in those with stimulation of saliva flow. Kato et al. *(31)* compared cocaine oral fluid concentrations in unstimulated to stimulated saliva. Unstimulated saliva collected by expectoration contained on average 5.2 (range 3.0–9.5) times as much cocaine compared with saliva collected under stimulated conditions (citric acid sour candy). The mean ratios of unstimulated to stimulated area under the concentration–time curve for benzoylegconine and ecgonine methyl ester were 6.0 (range 3.3–9.0) and 5.5 (range 2.5–8.8), respectively.

4. SAMPLE PREPARATION AND TESTING PROCEDURES

4.1. Sample Stability

There is a lack of data published on the subject of drug stability in biological matrices although it is widely recognized that drug levels can decrease over time. The degradation of drugs can be attributed to a number of factors including storage conditions, microbial action, and the adherence of the drug to the surfaces of the sample container. Current practices in oral fluid testing involve transport of the sample after collection for several days at varying temperatures and a significant length of time may pass between an initial presumptive screening test and a confirmation test required prior to the case going to court. Manufacturers of oral fluid collection devices have attempted to address this issue with the addition of buffers, to release the sample and drugs from the collection device, and preservatives, to stabilize the drugs in oral fluid.

4.2. Sample Pre-Treatment

Oral fluid samples diluted with buffer do not require any special pre-treatment for immunoassay screening and can also be analyzed directly by liquid chromatography–mass spectrometry (LC–MS). Solid-phase extraction (SPE) or liquid–liquid extraction (LLE) is recommended to reduce matrix interference. Subjecting the collected oral fluid samples to several freeze-thaw cycles followed by centrifugation can further reduce interference effects from mucins.

4.3. Screening Tests

Most screening tests for drugs-of-abuse are immunoassays. Antibodies used in immunoassays for detection of drugs in saliva must cross-react with the parent drug and lipophilic metabolites. For example, heroin and 6-acetylmorphine (6-AM), cocaine and ecgonine methyl ester, and Δ-9-tetrahydrocannabinol (Δ-9-THC) predominate in saliva. When drugs are leached into saliva from buccal depots such as is the case for smoked drugs, such as marijuana, parent drug and pyrolysis products will predominate in saliva. Immunoassays with cross-reactivity to free morphine and to 6-AM are most useful for the detection of opiates in oral fluid. Immunoassays that have been developed to detect the hydrophilic metabolites of drugs in urine will not be appropriate for saliva screening.

For effective evaluation of commercially available immunoassays for oral fluid, samples should be collected from both field and clinical trials and compared with a recognized reference method, for example, gas chromatography (GC)–MS *(32–34)*. Controlled administration studies are immensely important for establishing a greater understanding of pharmacokinetics *(35,36)* while the analysis of samples from known drug users provides important information on expected drug profiles *(37,38)*.

With advancements in technology, a number of researchers have investigated the use of LC–MS for multiple-drug screening as an alternative to immunoassays *(39–41)*. The main advantage of this technique is the use of a small sample volume coupled with the separation and identification power of the LC–MS. Immunoassays provide greater flexibility with respect to reduced sample handling and reduced costs for high-throughput testing laboratories.

4.4. POCT Testing (Immunoassays)

A number of on-site or point-of-care testing (POCT) devices have been introduced for screening drugs in oral fluid. The technology used in the majority of these devices is based on lateral flow of the oral fluid sample across a membrane impregnated with lines of labeled immobilized drug. These disposable cartridges are available for a range of drug groups including amphetamines, cocaine, opiates, PCP, and cannabinoids. The collection of oral fluid for on-site testing has provided a non-invasive alternative to blood or urine samples with application at the roadside, in clinical settings, and custody suits.

An early study conducted in 1999 to assess on-site tests by the roadside (ROSITA – Roadside) evaluated three devices: the Securetec Drugwipe, the Avitar OralScreen, and the Cozart® RapiScan *(42)*. The authors concluded that in general the sensitivity of the devices was not sufficient for low-drug concentrations seen for benzodiazepines and cannabinoids and the average total time for sampling and analysis was considered too long at 20 min (range 13–33).

A follow-up to this study, ROSITA 2, was started in 2003 to evaluate on-site devices by the roadside as before but to now include new and improved devices that had not been available for the initial study. Evaluation of the devices was extended to include use by police officers. The findings of the latest study were that six of the nine devices evaluated had an unacceptable number of failures (greater than 25%). In addition, the sensitivity and specificity of the devices did not meet the minimum specifications and no single device could be recommended for use at the roadside. The final report is available on the ROSITA website (www.ROSITA.org).

The Cozart® RapiScan (CRS) oral fluid drug testing system was the first fully integrated hand-held system for the detection of drugs. The CRS has been used successfully in police custody for testing individuals for use of opiates and cocaine who have been charged with a "trigger offence," for example, theft and drug offences *(11)*. The CRS has also been used successfully for monitoring heroin addicts *(43)* and detecting codeine and cocaine in controlled administration studies *(44,45)*.

4.5. Confirmation Testing and Tandem Mass Spectrometry

Chromatography coupled with MS is the technique of choice for confirming the presence of drugs in biological matrices. GC–MS and LC–MS are routinely used in analytical laboratories as they provide definitive identification of analytes of interest *(37–46)*.

The confirmation of drugs in oral fluid has proven a challenge to toxicologists due to limited sample volumes available for analysis and low cut-off concentrations. In response, a number of papers have been published recently employing tandem-MS to achieve greater sensitivity with small sample volumes. LC–MS–MS publications have been far greater with methods for amphetamines *(33)*, benzodiazepines *(39,40)*, and cannabinoids *(47)*, while GC–MS–MS methodologies predominantly involved cannabinoids *(48–50)* and opiates *(51,52)*.

The use of two-dimensional GC coupled to MS (GC–GC–MS) was reported for the sensitive analysis of 11-nor-delta-9-tetrahydrocannabinol-9-carboxylic acid (THC-COOH) in hair *(53)*. The application of this method to the analysis of drugs in oral fluid could provide an alternative approach for achieving the required sensitivity without the cost implications associated with tandem MS.

5. DRUGS DETECTED IN SALIVA/ORAL FLUID

5.1. Amphetamines

Amphetamine, methamphetamine, 3,4-methylenedioxymethamphetamine (MDMA), 3,4-methlyenedioxyamphetamine (MDA), and other amphetamine class drugs have saliva to plasma ratios (*S/P* ratios) of greater than one, ranging from 2 to 20. Saliva/plasma ratios of 2–4 during the elimination phase have been reported for amphetamine *(54,55)*, and for methamphetamine *(55)* and 3–20 for MDMA *(56)*. When 10 mg oral doses of amphetamine were given to volunteers, oral fluid concentrations of 10–60 ng/mL, peaking about 5 h after dosing (range 2–10 h depending on pH) were reported *(55)*. When amphetamine was administered to subjects as a racemic mixture, both D and L isomers were found in saliva *(55–57)*. After sustained release oral doses of methamphetamine, peak concentrations in oral fluid ranged from 14.5 to 33.9 ng/mL (10 mg) and 26.2 to 44.3 ng/mL (20 mg) and occurred within 2–12 h *(55)*. Kintz *(58)* reported *N*-methyl-1-(3,4-methylenedioxyphenyl)-2-butanamine (MBDB) and its metabolite (BDB) in oral fluid for 17 h after oral administration of 100 mg MBDB (concentration range 14–1045 ng/mL). The window of detection for methamphetamine in oral fluid has been reported from a minimum of 24 h after use *(55)* to as long as 50 h *(10,55,59)*.

5.2. Cannabinoids

Concentrations of the active ingredient of marijuana, Δ-9-THC, in oral fluid correlate well with Δ-9-THC concentrations in blood *(3,21,60)*. As little THC, and very little THCA, is secreted in saliva *(17,62)* this is most likely because of oral mucosa depots of cannabinoids as the source of Δ-9-THC in both oral fluid and serum *(21)*. Oral fluid concentrations of Δ-9-THC range from 1 to 400 ng/mL *(3,21,60,61)*. Teixeira et al. *(63)* reported that concentrations of THC in specimens obtained by direct spitting ranged from 50 to 6552 ng/mL while oral fluid retrieved from Salivette collectors by centrifugation from the same subjects at the same time contained from 0 to 134 ng/mL Δ-9-THC. Oral fluid specimens were positive at a cutoff of 1 ng/mL for an average of 5.7 h after smoking a 1.75% THC cigarette and for 8.8 h after a 3.55% Δ-9-THC cigarette *(21)*. Others report oral fluid to be positive after marijuana smoking for 2–10 h *(8–20)*. Niedbala et al. *(61)* reported that oral fluid specimens could be positive after passive exposure to marijuana smoke for up to 30 min after exposure. A follow-up study investigated the role of environmental contamination from the exposure of the collector to the cannabis smoke during the study *(49)*. Sample collection was carried out in a smoke-free environment in contrast to the initial study. All passive subjects were negative at screening/confirmation cut-off concentrations throughout the study. The authors concluded that the risk of a positive test for THC was virtually eliminated when specimens were collected in the absence of THC smoke.

5.3. Cocaine

Cocaine and cocaine metabolites are found in oral fluid. The *S/P* ratio for cocaine ranges from 0.5 to 10 *(10,15,64–68)*. The *S/P* ratio for benzoylecgonine ranges from 0.5 to 2.5 *(15,64,68)*, for ecgonine methyl ester from 2.3 to 5.1, for norcocaine 5.6 to 13.6, and for *para*-hydroxycocaine 2.4 to 10.8 *(68)*. Cocaethylene is found in saliva when ethanol is ingested concurrently with cocaine (*S/P* ratio ~1). After smoking cocaine, oral fluid concentrations of cocaine are elevated and the pyrolysis product of cocaine, anhydroecgonine methyl ester (AEME), is detected in oral fluid for up to 6 h but not in plasma *(15)*. However, the elevated saliva/plasma ratios of cocaine due to contamination of oral fluids clears within 2 h after smoking cocaine *(67)*.

Terminal half-lives in oral fluid are 7.9 h for cocaine, 9.2 h for benzoylecgonine, and 10 h for ecgonine methyl ester *(69)*. The detection time for cocaine in oral fluid ranges from 4 to 8 h after intravenous injection, 5 to 12 h after insufflation, and 12 h to 16 days after smoking cocaine *(15,67,68)*.

5.4. Opioids

Although all opioids may be subject to abuse, the principle illegal drug is heroin. The *S/P* ratios after intravenous administration of heroin are 0.1–1.9 for heroin, 0.7–7.2 for 6-AM, and 0.5–1.8 for morphine *(15)*. The presence of opiates in saliva after intravenous administration of heroin *(15,70)* and intramuscular administration of morphine *(71)* has been reported for up to 24 h. After 10 and 20 mg of heroin is administered intravenously, diacetylmorphine (6–20 ng/mL) is found in oral fluid for the first hour [terminal half life 8.5–582 min *(15)*] and then the primary metabolite of heroin, 6-AM (18–131 ng/mL) and morphine (10–16 ng/mL) are found for 4–8 h *(15,52,71–73)*. A longer window of detection would be expected where tolerant individuals use larger amounts of heroin.

Heroin may be smoked. Elevated *S/P* ratios (heroin 20–400, 6-AM 8–350, morphine 2–30) and higher drug concentrations (10–20,000 ng/mL heroin, 10–4000 ng/mL 6-AM, and 2–150 ng/mL morphine) may be found in oral fluid after smoking heroin *(15)*. After smoking, heroin was found in oral fluid for up to 24 h and 6-AM and morphine for up to 8 h *(15)*.

5.5. Phencyclidine

PCP has a theoretical *S/P* ratio of greater than 1 (pKa 9.4, plasma protein binding <10%). An *S/P* ratio of 1.5–3.0 was found in expectorated oral fluid after oral (1 mg) and intravenous (0.1 or 1 mg) administration of radiolabeled PCP to healthy male volunteers *(23)*. PCP and metabolites were detectable in saliva for 75 h. McCarron et al. *(24)* reported PCP concentrations ranging from 2 to 600 ng/mL in saliva samples from 100 patients suspected of PCP intoxication.

6. INTERPRETATION ISSUES

There are a number of factors that must be considered for effective interpretation of drugs in oral fluid. These include sample collection, dilution, and storage; the relationship between saliva and plasma for determining impairment; and the contribution of buccal drug depots and passive exposure.

Sample collection is the most important step in the process of oral fluid testing and can have a significant impact on the drugs detected. Collectors that cause excessive stimulation of saliva will reduce drug levels, and many sample collection devices have significant variability in the volume collected. Crouch *(74)* reported percentage volumes recovered between 18 and 83%, while Dickson et al. *(75)* reported excellent sample recoveries (CV < 10%) but highlighted the potential increase in error associated with incorporating a dilution step with the addition of preservative buffers. The use of a sample adequacy indicator will reduce the impact of the use of sample buffers.

A recent paper by Moore et al. *(53)* described losses for the Quantisal™ oral fluid collection device between 10 and 20% for Δ-9-THC when stored in the refrigerator and at room temperature, respectively. Crouch *(74)* reported losses of THC of greater than 40% with the Intercept® when stored under similar conditions. THC remained relatively stable in samples stored at –20 °C. Dickson et al. *(75)* investigated the effects of storage conditions (5 °C and room temperature) on drug stability and recovery using collection devices supplied by three different manufacturers. The study was restricted to investigating the recovery of THC, benzodiazepines, methamphetamine, and morphine. They concluded that storage at 5°C or room temperature had no significant effect on drug recoveries.

The variability of materials and preservatives used by different manufacturers will have a significant impact on the stability of drugs in oral fluid as demonstrated by the difference in THC stability for the two collection devices discussed previously. The impact of the collection device on drug stability must be assessed when interpreting an analytical result.

As it is the free lipophilic drug and drug metabolites that cross cell membranes such as the blood/brain barrier, and cause physiological effects, free drug in plasma, and related levels in saliva, may be correlated with drug effects. Through the use of the *S/P* ratio, there is the potential to relate drug levels in oral fluid to impairment, but this cannot be achieved with our current understanding of the mechanisms effecting drug levels in oral fluid. Saliva drug concentrations cannot be extrapolated to provide related levels in blood without knowing the saliva pH at the time of collection. The route of administration must also be considered as drug concentrations can be very high for several hours after smoking, snorting, or swallowing. The levels detected in oral fluid will not reflect those detected in plasma until the oral contamination has cleared.

The detection of marijuana in oral fluid is thought to be as a result of the deposition of cannabinoids in the oral mucosa following smoking of the drug. The measurement of Δ-9-THC, however, appears to follow the on-set and subsequent decline in physiological and pharmacological effects of marijuana.

7. ADVANTAGES AND DISADVANTAGES AS A DRUG TESTING MATRIX

The main advantage afforded by oral fluid is that the collection process is non-invasive and can be conducted under direct observation, therefore, minimizing the potential for adulteration of the sample. In addition, many collection devices have incorporated sample volume adequacy indicators ensuring that the required volume has been collected. It is extremely difficult to hold substances in the mouth in an attempt to adulterate or substitute the oral fluid sample, and this can be easily avoided by incorporating an oral cavity check as part of the collection procedure and/or a 10-min observation period prior to collection *(76)*. The collection can be carried out without the need for special facilities, so maintaining the dignity of the donor, and does not require same-sex collectors. The simplicity of the collection procedure has established oral fluid as a suitable alternative for near-patient testing and for investigating the role of drugs in impaired driving by sample collection and testing at the roadside.

The combination of low-drug concentrations and small sample volume presents a challenge to the analyst. Laboratories offering oral fluid tests have had to invest heavily in state-of-the-art instrumentation to achieve the required sensitivities. Another disadvantage associated with oral fluid testing is the short window of detection with many drugs being undetectable in oral fluid within 24 h. In the case of marijuana, Δ-9-tetrahydrocannabinol may be detected for only a few hours following smoking. The advantage of a short window of detection is the association with recent use and potential links to assessment of impairment. The inability to provide a sample on demand (comparable with a "shy bladder" for urine testing) is also a feature of oral fluid testing. Dry mouth is a common symptom associated with stress and some drugs will inhibit saliva secretions. Offering the donor a drink of water and waiting for 10 min before attempting another collection is recommended, but if the problem persists an alternate biological specimen should be collected.

8. FUTURE DEVELOPMENTS

The full potential of oral fluid has yet to be realized, and this presents a number of challenges to those already working with oral fluid and for those interested in incorporating oral fluid testing into their laboratory. At present, there are no recognized guidelines or quality standards associated with the analysis and interpretation of drugs in oral fluid. In the workplace testing field, draft guidelines are in circulation both in the USA (SAMHSA) and the United Kingdom (UK Steering Group for Workplace Drug Testing). The formalization of these guidelines will provide much needed guidance on best practice including cut-off concentrations and recommended collection procedures.

Testing laboratories worldwide recognize the importance of quality assurance in maintaining standards and work to both national (ABFT and NLCP) and international quality standards (ISO17025:2005). Including oral fluid into laboratory accreditation schemes can be difficult as the auditors do not have access to industry standards, and effective assessment of laboratory performance requires external quality assessment (EQA). Only recently have commercially quality control materials and proficiency testing schemes for oral fluid become available. One laboratory took the decision to commission an agency to design and instigate an in-house EQA scheme *(77)*.

Oral fluid testing has the potential to allow testing of drivers at the roadside and ultimately will provide an indication of an individuals' impairment due to drugs. Further research is required to optimize the benefits of using oral fluid in both clinical and criminal justice settings.

REFERENCES

1. Dvorchik B and Vessell E. Pharmacokinetic interpretation of data gathered during therapeutic drug monitoring. *Clin. Chem.* **22**(6):868–878; 1976.
2. Peel HW, Perrigo BJ and Mikhael NZ. Detection of drugs in saliva of impaired drivers. *J. Forensic Sci.* **29**(1):185–189; 1984.
3. Toennes S, Steinmeyer S, Maurer H, Moeller M and Kauert G. Screening for drugs of abuse in oral fluid–correlation of analysis results with serum in forensic cases. *J. Anal. Toxicol.* **29**(1):22–27; 2005.
4. Biermann T, Schwarze B, Zedler B and Betz P. On-site testing of illicit drugs: the use of the drug-testing device "Toxiquick." *Forensic Sci. Int.* **143**(1):21–25; 2004.
5. Samyn N, De Boeck G and Verstraete A. The use of oral fluid and sweat wipes for the detection of drugs of abuse in drivers. *J. Forensic Sci.* **47**(6):1380–1387; 2002.
6. Steinmeyer S, Ohr H, Maurer H and Moeller M. Practical aspects of roadside tests for administrative traffic offences in Germany. *Forensic Sci. Int.* **121**(1–2):33–36; 2001.
7. Kintz P, Crimele V and Ludes B. Detection of cannabis in oral fluid (saliva) and forehead wipes (sweat) from impaired drivers. *J. Anal. Toxicol.* **24**:557–561; 2000.
8. Idowu O and Caddy B. A review of the use of saliva in the forensic detection of drugs and other chemicals. *J. Forensic Sci. Soc.* **22**:123–135; 1982.
9. Yacoubian G, Wish E and Perez D. A comparison of saliva testing to urinalysis in an arrestee population. *J. Psychoactive Drugs* **33**(3):289–294; 2001.
10. Schramm W, Smith R, Craig P and Kidwell D. Drugs of abuse in saliva: a review. *J. Anal. Toxicol.* **16**:1–9; 1992.
11. Cooper G, Wilson L, Reid C, Main L and Hand C. Evaluation of the Cozart RapiScan drug test system for opiates and cocaine in oral fluid. *Forensic Sci. Int.* **150**:239–243; 2005.
12. Dawes C and Jenkins G. The effects of different stimuli on the composition of saliva in man. *J. Physiol.* **170**:86–100; 1964.
13. Malamud D. Guidelines for saliva nomenclature and collection. *Ann. N. Y. Acad. Sci.* **694**:xi–xii; 1993.
14. Haeckel R and Hanecke P. Application of saliva for drug monitoring an in vivo model for transmembrane transport. *Eur. J. Clin. Chem. Biochem.* **34**:171–191; 1996.
15. Jenkins A, Oyler J and Cone E. Comparison of heroin and cocaine concentration in saliva with concentrations in blood and plasma. *J. Anal. Toxicol.* **19**:359–374; 1995.
16. Jenkins A, Keenan R, Jenningfield J and Cone E. Pharmacokinetics and pharmacodynamics of smoked heroin. *J. Anal. Toxicol.* **18**:317–30; 1994.
17. Ohlsson A, Lindgren J, Andersson S, Agurell S, Gillespie H and Hollister L. Single-dose kinetics of deuterium-labeled cannabidiol in man after smoking and intravenous administration. *Biomed. Environ. Mass Spectrom.* **13**:77–83; 1986.

18. Cone E. Saliva testing for drugs of abuse. *Ann. N. Y. Acad. Sci.* **694**:91–127; 1993.
19. Niedbala R, Kardos K, Fritch D, Kardos S, Fries T, Waga J, Robb J and Cone E. Detection of marijuana use by oral fluid and urine analysis following single-dose administration of smoked and oral marijuana. *J. Anal. Toxicol.* **25**:289–303; 2001.
20. Maseda C, Hama K, Fukui Y, Matsubara K, Takahashi S and Akane A. Detection of Δ9-THC in saliva by capillary GC/ECD after marihuana smoking. *Forensic Sci. Int.*, **32**:259–266; 1986.
21. Huestis M and Cone E. Relationship of delta-9-tetrahydrocannabinol concentrations in oral fluid and plasma after controlled administration of smoked cannabis. *J. Anal. Toxicol.* **28**(6):394–399; 2004.
22. Cook C, Jeffcoat A, Hill J, Pugh D, Patetta P, Sadler B, White W, Perez-Reyes M. Pharmacokinetics of methamphetamine self-administered to human subjects by smoking S(+)methamphetamine hydrochloride. *Drug. Metab. Dispos.* **21**:717–723; 1993.
23. Cook C, Brine D, Jeffcoat A, Hill J, Wall M, Perez-Reyes M and DiGuiseppi S. Phencyclidine disposition after intravenous and oral doses. *Clin. Pharmacol. Ther.* **31**:625–634; 1982.
24. McCarron M, Walberg C, Soares J, Gross S and Baselt R. Detection of phency-clindine usage by radioimmunoassay of saliva. *J. Anal. Toxicol.* **8**:197–201; 1984.
25. ONeal C, Crouch D, Rollins D and Fatah A. The effects of collection method on oral fluid codeine concentrations. *J. Anal. Toxicol.* **24**:536–542; 2000.
26. Navarro M, Pichini S, Farre M, Ortuno J, Roset P, Segura J and de la Torre R. Usefulness of saliva for measurement of 3,4-methylenedioxymethamphetamine and its metabolites: correlation with plasma drug concentrations and effect of salivary pH. *Clin. Chem.* **47**:1788–1795; 2001.
27. Samyn N and Van Haeren C. On-site testing of saliva and sweat with Drugwipe and determination of concentration of drugs of abuse in saliva, plasma and urine of suspected users. *Int. J. Legal Med.* **113**:150–154; 2000.
28. Miles MV, Tennison MB, Greenwood RS. Intraindividual variability of carba-mazepaine, phenobarbital and phenytoin concentrations in saliva. *Ther. Drug Monit.* **13**:166–171; 1991.
29. Kintz P and Samyn N. Use of alternative specimens: drugs of abuse in saliva and doping agents in hair. *Ther. Drug Monit.* **24**(2):239–246; 2002.
30. Kauert G. Drogennachweis im Speichel vs. Serum. *Blutalkohol* **37**:76–83; 2000.
31. Kato K, Hillsgrove M, Weinhold L, Gorelick D, Darwin W and Cone E. Cocaine and metabolite excretion in saliva under stimulated and nonstimulated conditions. *J. Anal. Toxicol.* **17**:338–341; 1993.
32. Rana S, Moore C, Vincent M, Coulter C, Agrawal A and Soares J. Determination of propoxyphene in oral fluid. *J. Anal. Toxicol.* **30**(8):516–518; 2006.
33. Laloup M, Tilman G, Maes V, De Boeck G, Wallemacq P, Ramaekers J and Samyn N. Validation of an ELISA-based screening assay for the detection of amphetamine, MDMA and MDA in blood and oral fluid. *Forensic Sci. Int.* **153**(1):29–37; 2005.
34. Cooper G, Wilson L, Reid C, Baldwin D, Hand C and Spiehler V. Validation of an EIA microtiter plate assay for amphetamines in oral fluid. *Forensic Sci. Int.* **159** (2–3):104–112; 2006.

35. Barnes A, Kim I, Schepers R, Moolchan E, Wilson L, Cooper G, Reid C, Hand C and Huestis M. Sensitivity, specificity and efficiency in detecting opiate use in oral fluid with the Cozart® opiate microplate EIA and GC/MS following controlled codeine administration. *J. Anal. Toxicol.* **27**(7):402–406; 2003.
36. Kim I, Barnes A, Schepers R, Oyler J, Moolchan E, Wilson L, Cooper G, Reid C, Hand C and Huestis M. Sensitivity, specificity and efficiency in detecting cocaine use in oral fluid with the Cozart cocaine metabolite microplate EIA and GC/MS at proposed screening and confirmation cutoffs. *Clin. Chem.* **49**(9):1498–1503; 2003.
37. Cooper G, Wilson L, Reid C, Baldwin D, Hand C and Spiehler V. Analysis of cocaine and cocaine metabolites in oral fluid by Cozart microplate EIA and GC/MS. *J. Anal. Toxicol.* **28**(6): 498–503; 2004.
38. Cooper G, Wilson L, Reid C, Baldwin D, Hand C and Spiehler V. Comparison of GC/MS and EIA results for analysis of methadone in oral fluid. *J. Forensic Sci.* **150**(4):928–932; 2005.
39. Allen K, Azad R, Field H and Blake D. Replacement of immunoassay by LC tandem mass spectrometry for the routine measurement of drugs of abuse in oral fluid. *Ann. Clin. Biochem.* **42**(4):277–284; 2005.
40. Kintz P, Villain M, Concheiro M and Cirimele V. Screening and confirmatory method for benzodiazepines and hypnotics in oral fluid by LC-MS-MS. *Forensic Sci. Int.* **150**(2–3):213–220; 2005.
41. Wylie F, Torrance H, Anderson R and Oliver J. Drugs in oral fluid. Part I. Validation of an analytical procedure for licit and illicit drugs in oral fluid. *Forensic Sci. Int.* **150** (2–3):191–198; 2005.
42. Verstraete A and Puddu M. Evaluation of different roadside drug tests. *Rosita*, Roadside Testing Assessment, Rosita Consortium, Gent, Belgium, 167–232; 2001.
43. De Giovanni N, Fucci N, Chiarotti M. Scarlata S. Cozart RapiScan system: our experience with saliva tests. *J. Chromatogr. B* **773**(1):1–6; 2002.
44. Kolbrich E, Kim I, Barnes A, Moolchan E, Wilson L, Cooper G, Reid C, Baldwin D, Hand C and Huestis M. Cozart RapiScan Oral Fluid Drug Testing System: an evaluation of sensitivity, specificity, and efficiency for cocaine detection compared with ELISA and GC-MS following controlled cocaine administration. *J. Anal. Toxicol.* **27**(7):402–406; 2003.
45. Kacinko S, Barnes A, Kim I, Moolchan E, Wilson L, Cooper G, Reid C, Baldwin D, Hand C and Huestis M. Performance characteristics of the Cozart RapiScan Oral Fluid Drug Testing System for opiates in comparison to ELISA and GC/MS following controlled codeine administration. *Forensic Sci. Int.* **141**(1):41–48; 2004.
46. Scheidweiler K and Huestis M. A validated gas chromatographic-electron impact ionization mass spectrometric method for methylenedioxymethamphetamine (MDMA), methamphetamine and metabolites in oral fluid. *J. Chromatogr. B* **835**(1–2):90–99; 2006.
47. Laloup M, Del Mar Ramirez Fernandez M, Wood M, De Boeck G, Maes V and Samyn N. Correlation of Delta9-tetrahydrocannabinol concentrations determined by LC-MS-MS in oral fluid and plasma from impaired drivers and evaluation of the on-site Drager DrugTest. *Forensic Sci. Int.* **161**(2–3):175–179; 2006.
48. Day D, Kuntz D, Feldman M and Presley L. Detection of THCA in oral fluid by GC-MS-MS. *J. Anal. Toxicol.* **30**(9):645–650; 2006.

49. Niedbala S, Kardos K, Fritch D, Kunsman K, Blum K, Newland G, Waga J, Kurtz L, Bronsgeest M and Cone E. Passive cannabis smoke exposure and oral fluid testing. II. Two studies of extreme cannabis smoke exposure in a motor vehicle. *J. Anal. Toxicol.* **29**(7):607–615; 2005.

50. Teixeira H, Proenca P, Castanheira A, Santos S, Lopez-Rivadulla M, Corte-Real F, Marques E and Vieira D. Cannabis and driving: the use of LC-MS to detect delta-9-tetrahydrocannabinol (delta-9-THC) in oral fluid samples. *Forensic Sci. Int.* **146** (Supp.):S61–63; 2004.

51. Niedbala S, Kardos K, Waga J, Fritch D, Yeager L, Doddamane S and Schoener E. Laboratory analysis of remotely collected oral fluid specimens for opiates by immunoassay. *J. Anal. Toxicol.* **25**(5):310–315; 2001.

52. Presley L, Lehrer M, Seiter W, Hahn D, Rowland B, Smith M, Kardos K, Fritch D, Salamone S, Niedbala R and Cone E. High prevalence of 6-acetylmorphine in morphine positive oral fluid specimens. *Forensic Sci. Int.* **133**(1–2):22–25; 2003.

53. Moore C, Rana S, Coulter C, Feyerherm F and Prest H. Application of two-dimensional gas chromatography with electron capture chemical ionization mass spectrometry to the detection of 11-nor-Delta9-tetrahydrocannabinol-9-carboxylic acid (THC-COOH) in hair. *J. Anal. Toxicol.* **30**(3):171–177; 2006.

54. Wan S, Matin S and Azarnoff D. Kinetics, salivary excretion of amphetamine isomers and effect of urinary pH. *Clin. Pharmacol Ther.* **23**:585–590; 1978.

55. Schepers R, Oyler J, Joseph R Jr., Cone E, Moolchan E and Huestis M. Methamphetamine and amphetamine pharmacokinetics in oral fluid and plasma after controlled oral methamphetamine administration to human volunteers. *Clin. Chem.* **49**(1):121–132; 2003.

56. Samyn N, De Boeck G, Wood M, Lamers C, De Waard D, Brookhuis K, Verstraete A and Riedel W. Plasma, oral fluid and sweat wipe ecstasy concentrations in controlled and real life conditions. *Forensic Sci. Int.* **128**(1–2):90–97; 2002.

57. Cooper G, Peters F, Maurer H and Hand C. *Compliance of Individuals Prescribed Dexedrine® through Determination of Amphetamine Isomer Ratios in Oral Fluid.* Presented at the Society of Forensic Toxicologists Meeting, Dallas; 2004.

58. Kintz P. Excretion of MBDB and BDB in urine, saliva and sweat following single oral administration. *J. Anal. Toxicol.* **21**:570–575; 1997.

59. Suzuki S, Inoue T, Hori H and Inayama S. Analysis of methamphetamine in hair, nail, sweat and saliva by mass fragmentography. *J. Anal. Toxicol.* **13**:176–178; 1989.

60. Staub C. Chromatographic procedures for determination of cannabinoids in biological samples, with special attention to blood and alternative matrices like hair, saliva, sweat and meconium. *J. Chromatogr. B* **733**(1–2):119–126; 1999.

61. Niedbala S, Kardos K, Salamone S, Fritch D, Bronsgeest M and Cone E. Passive cannabis smoke exposure and oral fluid testing. *J. Anal. Toxicol.* **28**(7):546–552; 2004.

62. Perez-Reyes M. Marijuana smoking: factors that influence te bioaviailbility of tetrahydrocannabinol. In CN Chiang and RL Hawks, Eds, *Research Finding on Smoking of Abused Substances* . NIDA Research Monograph 99. Rockville, MD:42–62; 1990.

63. Teixeira H, Procena P, Verstraete A, Corte-Real F and Vieira D. Analysis of delta9-tetrahydrocannabinol in oral fluid samples using solid-phase extraction and high-

performance liquid chromatography-electrospray ionization mass spectrometry. *Forensic Sci. Int.* **150**(2–3):205–211; 2005.

64. Schramm W, Craig P, Smith R and Berger G. Cocaine and benzoylecgonine in saliva, serum and urine. *Clin. Chem.* **39**:481–487; 1993.

65. Cone E, Kumor K, Thompson L and Sherer M. Correlation of saliva cocaine levels with plasma levels and with pharmacologic effects after intravenous cocaine administration in human subjects. *J. Anal. Toxicol.* **12**:200–206; 1988.

66. Cone E and Weddington W (1989). Prolonged occurrence of cocaine in human saliva and urine after chronic use. *J. Anal. Toxicol.* **13**:65–68; 1989.

67. Cone E, Oyler J and Darwin W. Cocaine disposition in saliva following intravenous, intranasal and smoked administration. *J. Anal. Toxicol.* **21**, 465–475; 1997.

68. Jufer R, Wstadik A, Walsh S, Levine B and Cone J. Elimination of cocaine and metabolites in plasma, saliva and urine following repeated oral administration to human volunteers. *J. Anal. Toxicol.* **24**:467–477; 2000.

69. Moolchan E, Cone E, Wstadik A, Huestis M and Preston K. Cocaine and metabolite elimination patterns in chronic cocaine users during cessation: plasma and saliva analysis. *J. Anal. Toxicol.* **24**:458–466; 2000.

70. Gorodetzky C and Kullberg M. Validity of screening methods for drugs of abuse in biological fluids. II. Heroin in plasma and saliva. *Clin. Pharmacol. Ther.* **15**: 570–587; 1974.

71. Cone E. Testing human hair for drugs of abuse. I. Individual dose and time profiles of morphine and codeine in plasma, saliva, urine and beard compared to drug induced effects on pupils and behavior. *J. Anal. Toxicol.* **14**:1–7; 1990.

72. Cooper G, Wilson L, Reid C, Baldwin D, Hand C and Spiehler V. Analysis of opiates in oral fluid by Cozart microplate EIA and GC/MS. *Forensic Sci. Int.* **154** (2–3):240–246; 2005.

73. Moore L, Wicks J, Spiehler V and Holgate R. Gas chromatography /mass spectrometry confirmation of Cozart Rapiscan saliva methadone and opiates tests. *J. Anal. Toxicol.* **25**:520–524; 2001.

74. Crouch D. Oral fluid collection: the neglected variable in oral fluid testing. *Forensic Sci. Int.* **150**(2–3):165–173; 2005.

75. Dickinson S, Park A, Nolan S, Kenworthy S, Nicholson C, Midgley J, Pinfold R and Hampton S. The recovery of illicit drugs from oral fluid sampling devices. *Forensic Sci. Int.* **165**(1):78–84; 2007.

76. Jehanli A, Brannan S, Moore L, and Spiehler VR. Blind trials of an onsite saliva drug test. *J. Forensic Sci.* **46**(**5**): 1214–1220; 2001.

77. Clarke J and Wilson J. Proficiency testing (external quality assessment) of drug detection in oral fluid. *Forensic Sci. Int.* **150**:161–164; 2005.

Chapter 6

The Detection of Drugs in Sweat

Neil A. Fortner

Summary

Drug testing is commonly used as a deterrent to drug use in both the workplace and criminal justice systems. While the majority of these tests are conducted in urine, an increasing number of tests are being conducted using alternative biological specimens such as hair, oral fluid, blood, and sweat. The purpose of this chapter is to review the use of sweat as a biological matrix for the testing of drugs subject to abuse. This chapter will provide an overview of the structure of the skin, describe the composition of sweat, how the body produces sweat, examine the approaches used to collect sweat for analysis, provide an overview of the history of the detection of drugs in sweat, describe analytical approaches for the testing of sweat for the presence of drugs including procedures, and finally address the interpretation of drug test results in sweat.

Key Words: Drugs, sweat, sweat patch.

1. INTRODUCTION

In recent years, the drug testing community has seen a significant increase in the use of "alternative biological specimens" such as oral fluid, hair, and sweat for the detection of drugs subject to use and abuse. These alternative biological specimens offer different detection times and in most instances, significantly different metabolic profiles when compared to traditional urine testing. In addition, the fact that there continues to be issues with urine drug testing related to substitution, dilution, and specimen adulteration is the very reason that there has been an increase in the popularity and use of these

From: *Forensic Science and Medicine: Drug Testing in Alternate Biological Specimens*
Edited by: A. J. Jenkins © Humana Press, Totowa, NJ

alternative biological specimens for the detection of drugs subject to abuse. The purpose of this chapter will be to examine the use of sweat as one of these alternative biological specimens.

2. COMPOSITION OF THE SKIN

The skin is composed of two major layers: the epidermis and the dermis. The epidermis is the outermost layer of the skin and is composed of stratified epithelium. This layer varies between 75 and 150 μm in thickness over most of the body, except the palms of the hands and soles of the feet. The outer surface of the epidermis is called the stratum corneum. This layer acts as a barrier to restrain the passage of water and solutes in either direction across the skin *(1,2)*. Beneath the epidermis lies the dermis, the second major layer, and it is comprised of dense fibroelastic connective tissue. The dermis supports extensive vascular and nerve networks in addition to specialized excretory and secretory glands *(2)*.

2.1. Composition of Sweat

Moisture is lost from the skin by two distinct pathways. The first of these pathways appears to be caused by the diffusion of moisture through the dermal and epithelial layers *(3)*. Moisture lost through this mechanism is referred to as insensible sweat and results from the passive diffusion of volatiles, including water through the skin *(4)*. The rate of fluid loss through insensible sweat appears to depend on body location, ambient temperature, body temperature, and the relative humidity of the environment *(5)*. The second pathway for moisture lost is called sensible sweat and is most commonly referred to as sweat. This moisture is secreted from the eccrine glands *(5)* and the apocrine glands *(6)*. The innermost cells of the eccrine gland serve as secretory elements that then empty into either the lumen or the duct of the glands. The duct terminates at the surface of the skin *(7)*. There are a variety of triggers that stimulate the eccrine glands, which include but are not limited to exercise, stress (both mental and emotional), and thermal stress. The maximum rates of sensible sweat produced by the human body can be as high as 2 L/h in average subjects and as high as 4 L/h in trained athletes *(5)*. The amount secreted can be influenced by a variety of factors such as physical, thermal, and emotional stress *(8)*. The uneven distribution of sweat glands and the variability of these factors makes it difficult to systematically obtain sweat specimens.

Two other types of secretory glands common in human skin are the apocrine and sebaceous gland. Apocrine glands are located in the axillae, pubic, and mammary areas. They are not thought to have significant thermoregulatory use, and their role and mechanism in humans is unclear *(2)*. The sebaceous gland secretes a substance called sebum, that is primarily comprised of lipids. While these glands are located over most of the body, they do not appear to be involved in thermoregulation, and their function does not seem to be directly related to the production of sweat.

The mechanism by which drugs are incorporated into sweat is as yet unclear. Sweat originates from the blood/plasma, and its composition is largely determined by reabsorption and exchange mechanisms. As the pH of sweat may differ considerably from individual to individual, so may the composition. There is a strong correlation between the pKa of a drug and the amount found in the sweat.

3. THE COLLECTION OF SWEAT

A variety of methods have been employed for the collection of sweat. These approaches have included thermal stress followed by the collection of sweat using dry gauze covered with waterproof plastic *(9)* and the use of salt-impregnated pads *(10)*. Other sweat collection devices have included the use of cotton swabs *(11)*, polyvinyl shirts *(12)*, perspiration stains from clothing *(13)*, drug wipes *(14)*, and more elaborate devices such as a transcutaneous chemical collection device that incorporated a band-aid-like device with a water/gel matrix to absorb compounds and prevent back diffusion of these compounds into the skin *(15)*. Other devices have incorporated pilocarpine to stimulate localized sweat production *(16)*, and at least one device incorporated electrodes that dispensed pilocarpine for iontophoresis *(17)*. However, these devices had a variety of restrictions and limitations. Those that utilized a plastic covering *(9,12)* prevent the skin from breathing and significantly altered the physiology of the underlying epidermis. Other approaches *(11,12,14)* have limited use as they are primarily used to collect neat sweat or used as a surface wipe. Devices that utilize pilocarpine work well to stimulate the production of neat sweat but are limited in the amount of sweat they collect *(16,17)*.

The limitations observed with these sweat collection devices and approaches led to the development of a non-occlusive sweat collection device that permitted the evaporation of the trapped water content of the sweat. Eliminating the build up of this moisture minimizes skin irritation caused by moisture trapped against the skin. Because it would be possible to collect sweat components other than water over time, the potential to investigate drug metabolism and dose-dependent studies increased significantly.

In the late 1980s, Sudormed (Santa Ana, CA) developed a device known as the Sudormed Sweat Patch Specimen Container ™. This device was cleared by the US Food and Drug Administration (FDA) as a collection device in 1990 (Document Control No. K902442) and is comprised of three significant components. The first of these components is a polyurethane/adhesive layer consisting of 3M's Tegaderm™ 1625 transparent wound dressing. This dressing has been used since 1980 as a medical dressing and is widely used in hospitals and clinics to secure and protect catheters and intravenous lines. The adhesive used by 3M (Minneapolis, MN) is described as a hypoallergenic, water-resistant adhesive and has a tamper-evident characteristic. The adhesive used by 3M infiltrates the exfoliated stratum corneum cells. When the adhesive is removed from the skin, these cells adhere to the adhesive and prevent the adhesive from resticking. Consequently, this provides a tamper-evident mechanism to diminish the removal and reapplication of the sweat patch. A unique identification number is printed on the 3M Tegaderm™, which aids in the identification of the sweat patch and is a significant deterrent toward substitution. The polyurethane/adhesive layer is approximately 6 cm wide, 7 cm long, and 0.025 mm in thickness.

The second component of the sweat patch is the release liner. This is a very thin medical grade tissue paper that allows the release of the collection pad from the adhesive after patch wear. The release liner is approximately 3 cm wide, 5 cm long, and 0.003 mm thick. Without this release liner, the collection pad will adhere to the adhesive. After wearing the patch, the collection pad is retained and the release liner along with the Tegaderm™ polyurethane/adhesive layer is discarded.

The third component of the sweat patch is the collection pad. It is composed of medical grade cellulose supplied by Ahlstrom Filtration (Mt. Holly Springs, PA). The collection pad is approximately 3 cm wide and 5 cm long. It is approximately 0.7 mm thick and is the main component for the collection of the non-volatile components of sweat. The collection pad absorbs a minimum of 300 µl of insensible perspiration at ambient temperature. Studies involving controlled dosing have demonstrated that under these conditions, the sweat patch must be worn for a minimum of 24 h to collect adequate drugs for analysis *(18)*. This wear time may be shortened if other factors such as exercise are present that increase sweat production.

The sweat patch can be worn on the upper arm, lower rib cage area, and the upper back. Because of the aggressive nature of the adhesive, care should be taken to ensure that the application area is not subject to vast amounts of flexing. The upper arm is the most common area for sweat patch application as it easily presents itself, and the bicep can be flexed prior to the sweat patch application to reduce irritation caused by tension between the skin and the adhesive. Prior to application of the sweat patch, the skin should be cleaned using two isopropanol wipes. Care must be taken to allow the alcohol to completely evaporate, otherwise a skin irritation could develop due to isopropanol trapped beneath the sweat patch.

In 1992, the worldwide marketing and distribution rights for the sweat patch for the detection of drugs and alcohol were purchased by PharmChem, Inc. (Fort Worth, TX). The name PharmChek™ was chosen for this application of the Sudormed Sweat Patch Specimen Container (Fig. 1). Later that same year, Sudormed submitted a pre-market notification 510(k) to the FDA for the use of the sweat patch for the detection of cocaine (COC) and metabolites in sweat. This pre-market notification focused on the detection of COC and metabolites using gas chromatography–mass spectrometry (GC/MS). The FDA subsequently notified Sudormed that this COC submission could not be approved without an approved method for screening for COC in sweat. The FDA requested that a separate 510(k) submission for a screening assay for the detection of COC be submitted along with data to support the use of the sweat patch as a collection device. In addition, the FDA requested that Sudormed provide similar separate 510(k) submissions for screening and confirmation for each of the other drugs [opiates, phencyclidine (PCP), amphetamines, and marijuana] for which it intended to have the sweat patch approved as a collection device. To comply with the FDA's request, Sudormed met with a number of manufacturers of screening reagents to assess their ability to detect the presence of drugs in sweat. As a result of these discussions, SolarCare Technologies Corporation (STC) in Bethlehem, PA, agreed to modify its enzyme-linked immunoassay microplate screening assay (ELISA) to detect the presence of drugs in sweat. Sudormed and SolarCare Technologies subsequently submitted co-dependent 510(k) documents for the detection of COC, opiates, PCP, amphetamines, and marijuana by the end of 1993. By the end

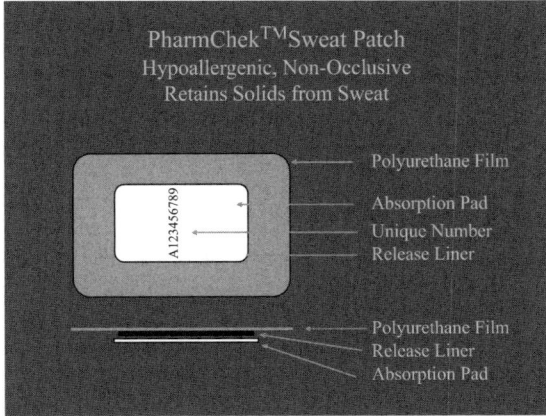

Fig. 1. Diagram of a sweat patch.

of 1996, all five submissions had been cleared by the FDA. STC underwent a merger in the late 1990s and is now know as OraSure Technologies.

4. THE DETECTION OF DRUGS IN SWEAT

There have been many studies designed to detect the presence of alcohol and other drugs in sweat. Pawan et al. *(12)* studied alcohol levels using polyvinyl shirts to collect sweat. Vree et al. *(19)* administered controlled oral doses of amphetamines and measured the amphetamine levels in sweat. This study indicated that the elimination of amphetamines in sweat was largely independent of the sweat pH. Ishiyama et al. *(11)* determined that methamphetamine was excreted by sweat glands and that the analysis of sweat would be a valuable tool in forensic practices. Henderson et al. *(20)* measured methadone and metabolites in both the sweat and urine of patients that were in methadone treatment programs, primarily to determine the optimal doses of methadone. The data suggested that the drug levels in sweat could not be used to determine optimal doses of methadone but indicated that sweat may be a significant route of elimination for drugs. Cook et al. *(21)* studied the detection of ^{3}H PCP in sweat by means of filter papers placed on the arm below the axilla. They determined that significant levels of radioactivity were present in the filter papers up to 54 h following drug administration. Parnas et al. *(22)* studied the excretion in sweat of phenytoin, phenobarbitone, and carbamazepine using an occlusive bandage consisting of one to three layers of filter paper covered by sampling padding. This study indicated that while all three drugs were found to be present in sweat, the phenytoin sweat concentrations was found to correspond to the free fraction in plasma and was independent of sweat flow. Phenobarbitone sweat concentrations were found to increase with increasing sweat flow. This data suggested that drug level monitoring, under changing climatic conditions, may be of clinical significance.

There have been numerous studies that have entailed controlled dosing studies for COC, at least one study involving a controlled dose for heroin, at least one study involving a controlled does for codeine, and at least one study involving the administration of PCP.

Cone et al. *(18)* examined the effect the route of administration and dose on sweat COC levels. A 25 mg intravenous dose of COC produced sweat COC levels that ranged from 11.0 to 44.7 ng/patch. A 32 mg dose of COC administered intranasally produced sweat COC levels that ranged from 31.2 to 39.8 ng/patch. A 42 mg smoked dose produced sweat COC levels that ranged from 4.7 to 73.6 ng/patch. This study also demonstrated that COC appears in sweat within 1–2 h of administration and that levels peaked within 24 h. In addition, it was noted that COC was detectable in trace amounts following the administration of 1 mg of the drug. In all cases, COC was the major analyte excreted in sweat, although smaller amounts of ecgonine methyl ester (EME) and benzoylecgonine (BE) were present.

Cone et al. *(18)* also demonstrated that a 20 mg intravenous dose of heroin produced sweat heroin levels that ranged from 6.9 to 53.3 ng/patch. Heroin was the major analyte excreted in sweat in this study although smaller amounts of 6-acetyl morphine were present. Morphine was not detected in any of these samples.

Kacinko et al. *(23)* administered three subcutaneous doses of COC (75 mg/70 kg) over a 1-week period and observed levels that ranged from 3.8 to 84.7 ng/patch. In a similar study, three subcutaneous doses of COC (150 mg/70 kg) were administered over a 1-week period. In this study, COC levels were as high as 375 ng/patch. Uemura et al. *(24)* administered 210 mg/70 kg of COC intravenously to volunteers and observed sweat COC levels as high as 375 ng/patch. Of particular interest in this study was the observation that sweat patches placed on the back of the individuals showed sweat COC levels that were up to eight times higher that those patches placed on the shoulders.

Burnes et al. *(25)* administered either 50 or 126 mg of COC hydrochloride intranasally to volunteers. A 1-week interval separated dosing periods, and the order of the doses were counterbalanced. Sweat patch levels for the 50 mg dose ranged from 0 to 105 ng/mL (0 to 263 ng/patch) while individuals receiving the 126 mg dose demonstrated sweat patch COC levels that ranged from 0 to 140 ng/mL (0 to 315 ng/patch).

Schwilke et al. *(26)* administered three doses of codeine sulfate within a 7-day period. Those individuals who were administered the low does of 60 mg/70 kg demonstrated peak sweat patch codeine levels that ranged from 0 to 225 ng/patch. Those individuals that received the high dose of 120 mg/70 kg of codeine sulfate demonstrated sweat patch codeine levels that ranged from 0 to 96 ng/patch.

Cook et al. *(21)* administered 0.5 mg of a deuterated PCP (D_5) to individuals intravenously using a bolus approach. The analysis was conducted using RIA coupled with HPLC confirmation and indicated a concentration of PCP D_5 in sweat of 5.9 ng/mL. In this study, the sweat was not collected using a sweat patch but was collected as neat dripped off the upper body of one of the subjects who was exercising vigorously on an exer-cycle.

5. SPECIMEN TESTING

5.1. Sweat Patch Extraction

Upon arrival at the testing facility, the sweat patch must be reconstituted to extract the drugs off of the collection pad and suspend them in an appropriate matrix using the process described by Spiehler et al. *(27)*. The sweat patch was placed in a 6-mL polypropylene specimen vial with screw cap. The sweat patch reconstitution is accomplished by adding 2.5 mL of a 25:75 0.25 M sodium acetate buffer : methanol solution to each sweat patch specimen. Each vial is capped and shaken on a horizontal shaker for 30 min at 2400 rpm. Following this extraction, the tube may be centrifuged to force the collection pad to the bottom, or a device similar to a serum separator may be used to force the collection pad to the bottom of the tube. Other investigators have used a similar approach but have used between 4 and 6 mL of the sodium acetate buffer : methanol solution to reconstitute the sweat patch *(18,23,26)* to allow for a greater volume of eluate for subsequent analysis. Previously, the results for sweat patch analysis have been reported in ng/mL units *(25)* using 2.5 mL as described above to reconstitute the sweat patch. The current approach is to present the sweat patch drug concentrations in a ng/patch format.

The current testing levels used by the industry for sweat testing are listed below. With the exception of PCP, all testing levels were established using receiver operating characteristic approaches as described by Spiehler et al. *(27)*. Testing levels in parenthesis are levels proposed by the US Department of Health and Human Services, Substance Abuse and Mental Health Services Administration *(28)*.

Testing Levels in Sweat

Screening levels		Confirmation levels	
Amphetamines	25 *(25)*	Amphetamine	25 *(25)*
		Methamphetamine	25^b $(25)^b$
		MDMA[a]	*(25)*
		MDA[a]	*(25)*
		MDEA[a]	*(25)*
Cocaine	25 *(25)*	Cocaine	25^b *(25)*
		Benzoylecgonine	25 *(25)*
Cannabinoids	3.75 *(4)*	Δ9 THC	1.25 *(1)*
Opiates	25 *(25)*	Morphine	25 *(25)*
		Heroin	25^c
		6-Acetyl Morphine	25 *(25)*
		Codeine	25 *(25)*
Phencyclidine	18.75 *(20)*	Phencyclidine	18.75 *(20)*

All levels have been converted to ng/patch. Parentheses indicates proposed DHHS testing levels.
[a]Indicates proposed DHHS additional drugs.
[b]Indicates metabolite requirement for reporting a positive drug result.
[c]Heroin missing from proposed DHHS regulations.

Currently, sweat patch drug results are expressed as ng/mL of reconstituted acetate buffer/methanol. The proposed SAMHSA guidelines express the drug levels as ng/patch. For example, using this approach, the cutoff for amphetamines would be 25 ng/patch (10 ng/mL × 2.5 mL of reconstitution matrix).

5.2. Screening

The OraSure Drugs-Of-Abuse Assays are micro-plate competitive immunoassays for the determination of drugs-of-abuse or their metabolites in PharmChek™ sweat patch eluate. Sample or standard is added to microtiter wells along with enzyme-labeled hapten conjugate. There is a competition to bind the antibody fixed onto the well. The wells are washed six times with deionized water to remove excess enzyme conjugate and drug once the competition for the antibody coated on the surface of the microplate well is complete. A substrate (3,3″,5,5″-tetramethylbenzidine) is added and incubated for an additional 30 min. Following this incubation, the reaction is stopped using 100 µL of 2 N sulfuric acid. The microplates are then read at two wavelengths 450 and 630 nm using a microplate reader. The readings at 630 nm are used to negate any wavelength distortions due to scratches or fingerprints on the microplates. The results of the 630 nm reading are subtracted from the results of the 450 nm reading. These absorbance reading are inversely proportional to the amount of drug or drug metabolite(s) present in the sample or calibrator/control.

Because EIA assays are based on the principle of an antibody recognizing a drug or class of drugs, metabolites or different members of a drug class cross-react with antibody to varying degrees. Therefore, EIA assays are used as qualitative procedures. A positive test result indicates the presence of the drug/drug class at or above the stated cutoff concentration. Specimens that produce an absorbance reading less than or equal to the cutoff value established by the calibration procedure are designated as an initial test "positive." Specimens that produce absorbance readings that are greater than or equal to the cutoff calibrator value are designated as "negative." Initial results are sent to the Data Review Department for evaluation. Specimens testing positive are sent to GC/MS for confirmation analysis. Specimens testing negative are not subjected to further analysis.

5.3. Confirmation

After a sample has been screened positive by EIA for a drug or drug group, GC/MS is performed on another aliquot of the specimen. The purpose of the additional testing is threefold: First, for those cases in which the EIA assays tests for a general class of drugs, GC/MS identifies the specific drug(s) that are present. Second, GC/MS quantifies the drug to compare each specimen to an established cutoff concentration, below which the specimen will be reported as negative. Third, GC/MS confirms all positive EIA results with a technology that is chemically different from EIA. In many cases, the drug is derivatized to enhance its chromatographic properties. Following extraction and derivatization, if applicable, the specimen is injected into a GC equipped with a fused silica capillary column and coupled to an electron impact (EI) mass spectrometer. Because of the lower confirmation levels for cannabinoids, negative chemical ionization (NCI) or coupled mass spectrometry (MS/MS) is required. Because the reconstitution matrix is an acetate buffer and methanol solution, confirmation using LC/MS/MS may be a more attractive approach than conventional GC/MS or GC/MS/MS.

Each analysis incorporates an internal standard that is added to the specimen. The internal standard is selected on the basis of chemical similarity to the drug(s) to be tested. In many cases, the internal standard is a deuterated derivative of the drug(s) to be analyzed. The internal standard is selected so as to extract, derivatize, and chromatograph in a manner similar to the analyte. It should also yield spectral masses that are close to, but different from, those of the analyte.

As they flow through the capillary column, constituent drugs are separated from interference and other drugs on the basis of their volatility and varying affinities to the column. As the effluent enters the mass spectrometer, it is subjected to an electron beam, which fragments the molecules, creating a "fingerprint" pattern unique to the drug(s) tested. The response of three ion fragments for each analyte and two ion fragments for each internal standard is monitored and the area measured using selected ion monitoring (SIM) mode. The compound retention time, the presence of the three-analyte ions, and their abundance ratios enable the conclusive identification of drugs in the unknown samples as compared to the standards. Quantitation is achieved by comparing the ratio of the area of the base peak of the analyte to the area of the base peak of the internal standard in a specimen to the analogous ratio derived from the calibration standard. The quotient of that calculation is multiplied by the known concentration of the calibration standard to determine the quantitative value of the unknown.

5.4. Amphetamines

Amphetamine and methamphetamine are eluted off the sweat patch using a methanol/acetate buffer. Amphetamine and methamphetamine are extracted, with deuterated analogs as internal standards, from alkalinized patch extracts. They are back-extracted into acid and alkalinized. Finally, they are reextracted into *n*-butyl chloride that is treated with 4-carbethoxyhexafluorobutyryl chloride (CB) to form the acyl derivatives. They are identified and quantitated by capillary GC EI three-ion SIM MS. No compounds have been identified which interfere with this assay although several other primary and secondary amines may be identified by this procedure.

Methamphetamine is metabolized into amphetamine; however, little amphetamine is detected in sweat. Methamphetamine can be detected in sweat using the patch in as short a time as 4 h. Levels in the patch are dose related but vary considerably between individuals.

5.5. *Cannabinoids*

Δ-9-Tetrahydrocannabinol (THC) is eluted off the sweat patch using a 75/25 methanol/acetate buffer combination. The mixture is diluted with an aqueous buffer, and the THC is extracted using a solid phase column. The extracted THC is derivatized using trifluoroacetic anhydride and analyzed by GC/MS using NCI. Quantitation is achieved using a deuterated internal standard.

5.6. *Cocaine*

COC and its metabolites, BE, and EME are eluted off the sweat patch using a methanol/acetate buffer mixture (75:25). The drugs are extracted from the mixture using a solid-phase column. The BE and EME are derivatized using HFIP and PFPA and analyzed by GC/MS using SIM. COC is analyzed in this same chromatographic assay in its underivatized form. Quantitation is achieved using deuterated internal standards.

COC and its two major metabolites, BE and EME, can be detected in sweat as early as 2 h after a single dose. The levels in the patch will reach a maximum between 2 and 3 days. Continued use of COC will result in increasing levels in the patch.

5.7. *Opiates*

Codeine, morphine, 6-acetylmorphine (6-AM) and heroin are eluted off the sweat patch using a 75/25 methanol/acetate buffer mixture. The eluate is made alkaline, and the drugs are extracted into an organic solvent. The drugs are back-extracted into acid that is alkalinized and then reextracted into organic solvent that is evaporated to dryness. The residue is treated with BSTFA and analyzed by GC/MS using SIM. Heroin is analyzed in its underivatized form. Quantitation is achieved using deuterated internal standards.

Table 1
PharmChem Personal Communication

Drug	Extraction	Derivative	Method	Ions	LOD/LOQ
Amphetamine	Liq/liq	4-CB	GC/MS EI	248,266,294	5/5
Methamphetamine	Liq/liq	4-CB	GC/MS EI	262,280,308	5/5
Cannabinoids	Solid phase	TFAA	GC/MS NCI	410.3, 413.3	0.5/0.5
Cocaine	Solid phase	HFIP/PFPA	GC/MS EI	182,272,303	5/5
Benzoylecgonine	Solid phase	HFIP/PFPA	GC/MS EI	318,334,439	5/5
Codeine	Solid phase	N/A	GC/MS EI	371,372,343	7.5/7.5
Morphine	Solid phase	BSTFA	GC/MS EI	429,414,430	10/12.5
6-MAM	Solid phase	BSTFA	GC/MS EI	399,340,287	7.5/7.5
Heroin	Solid phase	BSTFA	GC/MS EI	327,268,310	7.5/10
Phencyclidine	Liq/liq	None	GC/MS EI	200,242,243	2.2/2.5

4-CB, 4-carbethoxyhexafluorobutyryl chloride; EI, electron impact; GC/MS, gas chromatography–mass spectrometry; NCI, negative chemical ionization.

5.8. Phencyclidine

PCP is eluted off the sweat patch using a 75/25 methanol/acetate buffer combination. The eluate is made alkaline, and PCP is extracted into an organic solvent. The drug is back-extracted into acid that is then alkalinized. Finally, PCP is extracted into a small volume of chloroform : isoamyl alcohol (9:1) and is analyzed by GC/MS using SIM. Quantitation is achieved using deuterated internal standard.

At present, there is only one laboratory providing sweat patch testing services. PharmChem Inc. located in Fort Worth, TX, owns the world-wide marketing rights to the PharmChek™ Sweat Patch. Analytical parameters for sweat patch testing are summarized in Table 1 and were provided by PharmChem Inc (PharmChem Inc. *Standard Operating Procedures for the Testing of Sweat Patches*. Haltom City, TX. Personal Communication).

6. Interpretation of Results

Numerous studies *(11,13,14,18–22,29)* have confirmed that the predominant drug detected in sweat is the parent. Following administration, drugs may be detected in sweat in as little as 2 h *(18)*. These controlled dose studies concluded that a sweat patch must be worn for a minimum of 24 h to collect sufficient drug for analysis *(18)*. In addition, it appears that drugs in sweat lag behind the detection of the same drug in urine by 1–2 days. The sweat patch is a collection device and serves as a reservoir for drugs and other compounds secreted in sweat. Consequently, it cannot detect multiple uses of drugs that may have occurred while the sweat patch was worn and currently cannot detect what route of administration was used. Little research has been conducted regarding the lowest dose required to obtain a positive sweat patch result. Cone et al. *(18)* has reported that COC was detectable in the sweat patch in trace amounts following the administration of 1 mg of COC.

Concern has been raised regarding the susceptibility of the sweat patch to contamination from the incomplete cleaning of the skin with the isopropanol wipes *(30)*. A subsequent study concerning the incomplete cleaning of the skin considered the efficiency of the two isopropanol wipes at removing 1000 ng of methamphetamine and COC applied to the skin such that no drug was subsequently transferred to the sweat patch during normal wear. The authors also looked at the effects of a 5% soap solution followed by water and a single isopropanol wipe at removing up to 1000 ng of methamphetamine and COC from the skin *(31)*. They concluded that cleansing with soap and isopropanol was more effective at removing the residual drug from the skin than the two isopropanol wipes alone and that there was a relationship between the dose applied to the skin and the residual drug detected on the patch. Although the parent drug was sometimes detected following these treatments, neither the drug nor the metabolite was detected in the same patch. Consequently, under the existing procedures and the proposed SAMHSA guidelines, none of these sweat patches would have been reported as positive for either COC or methamphetamine, regardless of the cleaning procedure used.

Concern has been raised regarding the susceptibility of the sweat patch to vapor phase contamination *(30)*. This concern arose out of several criminal justice court cases in which the defendants claimed that they had not used COC and that their positive sweat patch test results must be due to the COC vapor resulting from other individuals in the household smoking crack COC. Kidwell et al. *(30)* designed a series of laboratory experiments to test the susceptibility of the sweat patch to environmental contamination using COC, methamphetamine, and heroin. These investigators concluded that the sweat patch was resilient to external vapor phase contamination if the sweat patch and the external surface of the patch were kept dry. However, if the pad was moistened with basic buffer, water, or artificial sweat, both COC and methamphetamine could be detected in the absorbent pad. The authors acknowledged that this project was primarily undertaken for mechanistic studies rather than reflecting a realistic real-life contamination scenario as the amounts of drug vapor used were large. A subsequent study was conducted by Crouch et al. *(32)* to determine whether the sweat patch absorbent pad was susceptible to external contamination (in vitro) when known amount of drug were volatilized in the presence of the sweat patch. The absorbent pad was left dry, moistened with artificial sweat, harvested drug-free sweat, or water, covered with the polyurethane Tegaderm™

material, and subjected to known amounts of volatilized drug *(32)*. The authors *(32)* concluded that the polyurethane membrane of the sweat patch serves as an effective barrier to vaporized COC and methamphetamine. Under a variety of physiologically relevant conditions, the sweat patch absorbent pad remained drug free. It would require gross exposure of the patch (equivalent to volatizing 200 g of COC and 100 g of methamphetamine) in a 20' × 20' × 8'-sized room and remaining in this room for at least 1 h to duplicate these experimental conditions *(32)* such that either COC or methamphetamine could be detected in the sweat patch. These conditions are certainly not representative of normal environmental conditions. In addition, while these extreme conditions did demonstrate the presence of either COC or methamphetamine on the absorbent pad, no BE or amphetamine were detected. Therefore, under the existing procedures and the proposed SAMHSA guidelines, none of these patches would have been reported as positive.

7. ADVANTAGES AND DISADVANTAGES

Sweat is an easy specimen to collect, with several methodologies available for use. Most drugs-of-abuse may be extracted from the sweat using current testing methodologies. Screening techniques for urine are generally not appropriate for sweat due to low-drug concentrations detected in sweat and the incorrect compound targeted. One study *(33)* demonstrated in a prison setting that wearing the sweat patch proved to be a deterrent to drug use and provides a means to monitor individuals in a correction or probation/parole setting for an extended period of time. Disadvantages include the need to wear the patch for a minimum of 24 h to achieve measurable drug concentrations. The sweat patch device appears to be resistant to substitution and contamination under realistic "real-life" conditions.

Although the concentration of drugs measured in sweat appear to be dose dependent, future research should focus on determining the factors contributing to the variability in concentrations. Other drugs-of-abuse such as hydrocodone and oxycodone should also be studied.

REFERENCES

1. Potts, R. O. and Guy, R. H., The predictions of percutaneous penetration: a mechanistic model, in *Dermal and Transdermal Drug Delivery*, Gurney, R., and Teubner, A. Eds., CRC Press, Boca Raton Florida, 1993.
2. Odland, G. F., Structure of the skin, in *Physiology, Biochemistry, and Molecular Biology of the Skin*, Lowell A. Goldsmith M.D., Oxford University Press, New York, 3–17, 1991.
3. List, C. F., Physiology of sweating, *Ann. Rev. Physiol.*, 10, 387–400, 1948.

4. Bucks, D. A., Maibach, H. I., and Guy, R. H., Mass balance and dose accountability in percutaneous absorption studies, *Pharm. Res.*, 5(55), 313–315, 1988.

5. Randall, W. C., Special review: the physiology of sweating, *Am. J. Physiol. Med.*, 32, 292, 1953.

6. Cleary, G. W., The first two decades of transdermal drug delivery systems and a peek into 1990's, in *Dermal and Transdermal Drug Delivery*, Gurney, R., Ed., Wissenchatlliche, Stuttgart, Germany, 1993, 13–32.

7. Sata, K. and Date, F., Individual variations in structure and function of human eccrine sweat gland, *Am. J. Physiol.*, 245(2), R203–R208, 1983.

8. Randall, W. C., Quantitation and regional distribution of sweat glands in man, *J. Clin. Invest.*, 25, 761, 1946.

9. Darling, R. C., diSant' Agneses, P. A., Perera, G. A., and Andersen, D. A., Electrolyte abnormalities of sweat in fibrocystic disease of pancreas. *Am. J. Med. Sci.*, 225, 67, 1953.

10. Philips, M., An improved adhesive patch for long-term collection of sweat, *Biomater. Med. Devices. Artif. Organs*, 8(1), 13–21, 1980.

11. Ishiyama, I., Nagai, T. O., Nagai, T. A., Komuro, E., Momose, T., and Akimori, N., The significance of drug analysis of sweat in respect to rapid screening for drug abuse, *Z. Rechtsmed.*, 82, 251–256, 1979.

12. Pawan, G. L. S. and Grice, K., Distribution of alcohol in urine and sweat after drinking. *Lancet*, 2, 1016, 1968.

13. Smith, F. P. and Pomposini, D. A., Detection of phenobarbital in bloodstains, semen, seminal stains, saliva, saliva stains, perspiration stains, and hair. *J. Forensic Sci.*, 26(3), 582–586, 1981.

14. Kidwell, D. A., Blanco, M. A., and Smith, F. P., Cocaine detections in a university population by hair analysis and skin swab testing. *Forensic Sci. Int.*, 84, 75–86, 1997.

15. Conner, D., Millora, E., Zamani, N., Almirez, R., Rhyne-Kirsch, P., and Peck, C., Transcutaneous chemical collection of caffeine in normal subjects; relationships to area under the plasma concentration-time curve and sweat production, *J. Invest. Dermatol.*, 2, 186–189, 1990.

16. LeGrys, V. A., Sweat testing, *Clin. Chem.*, News, April, 1992.

17. Warwick, W. J., Huang, N. N., Waring, W. W., Cherian, A. G., Brown, I., Stejskal-Lorenz, E., Yeung, W. H., Duhon, G., Hill, J. G., and Strominger, D., Evaluation of a cystic fibrosis screening system incorporating a miniature sweat stimulator and disposable chloride sensor. *Clin. Chem.*, 32(5) 850–853, 1986.

18. Cone, E. J., Hillsgrove, M., Jenkins, A., Keenan, R. M., and Darwin, W. D., Sweat testing for heroin, cocaine, and metabolites. *J. Anal. Toxicol.*, 18, 298–305, 1994.

19. Vree, T. B., Muskens, J. M., and Van Russon, J. M., Excretion of amphetamines in sweat, *Arch. Pharmacodyn.*, 199, 311–317, 1972.

20. Henderson, G. L. and Wilson, B. K., Excretion of methadone and metabolites in human sweat, *Res. Commun. Chem. Pathol. Pharmacol.*, 5, 1–8, 1973.

21. Cook, C. E., Brine, D. R., Jeffcoat, A. R., Hill, J. M., Wall, M. E., Perez-Reyes, M., and DiGuiseppi, S. R., Phencyclidine disposition after intravenous and oral doses, *Clin. Pharmacol. Ther.*, 31, 625–634, 1982.

22. Parnas, J., Flachs, H., Gram, L., and Wurtz-Jorgensen, A., Excretion of antiepileptic drugs in sweat. *Acta Neruol. Scand.*, 58, 197–207, 1978.
23. Kacinko, S. L., Barnes, A. J., Schwilke, E. W., Cone, E. J., Moolchan, E. T., and Huestis, M. A. Disposition of cocaine and its metabolites in human sweat after controlled cocaine administration., *Clin. Chem.*, 51, 11, 2085–2094, 2005.
24. Uemura, N., Nath, R. P., Harkey, M. R., Henderson, G. L., Mendelson, J., and Jones, R. T. Cocaine levels in sweat collections patches vary by location of patch placement and decline over time. *J. Anal. Toxicol.*, 28(4), 253–259, 2004.
25. Burnes, M and Baselt, R. C., Monitoring drug use with a sweat patch: an experiment with cocaine. *J. Anal. Toxicol.*, 19, 41–48, 1995.
26. Schwilke, E. W., Barnes, A. J., Kacinko, S. L., Cone, E. J., Moolchan, E. T., and Huestis, M. A. Opoid disposition I human sweat after controlled oral codeine administration. *Clin. Chem.*, 52, 8, 1539–1545, 2006.
27. Spiehler, V., Fay, J., Fogerson, R., Schoendorfer, D. and Niedbala, R. S., Enzyme immunoassay validation for qualitative detection of cocaine in sweat, *Clin. Chem.*, 42, 1, 34–38, 1996.
28. Department of Health and Human Services; Substance Abuse and Mental Health Services Administration; Mandatory Guidelines of Federal Workplace Drug Testing Programs; Notice of Proposed Revisions. Federal Register Doc 04–7984.
29. Goldberger, B. A., Darwin, W. D., Grant, T. M., Allen, A. C., Caplan, Y. H., and Cone, E. J., Measurement of heroin and it's metabolites by isotope-dilution electron-impact mass spectrometry. *Clin. Chem.*, 39, 870–875, 1993.
30. Kidwell, D. and Smith, F., *Susceptibility of the PharmChek Drugs of Abuse Patch to Environmental Contamination.* Naval Research Laboratory, Washington, DC, 1999.
31. Crouch, D. J., Metcalf, C. L., and Slawson, M. H., An assessment of the effectiveness of the PharmChek™ sweat patch skin cleaning procedure. *Bull. Inter. Assoc. Forensic Toxicol.*, 32(2), 5–8, 2002.
32. Crouch, D. J., Metcalf, C. L., Slawson, M. H., and Baudy, J., An assessment of the potential for vapor phase contamination of the PharmChek™ sweat patch. *Bull. Inter. Assoc. Forensic Toxicol.*, 32(3), 7–10, 2002.
33. Sunshine, I. and Sutliff, J. P., Sweat it out, in *Handbook of Analytical Drug Monitoring and Toxicology*, CRC Press, Inc, Boca Raton, Florida, 253–226, 1997.

Chapter 7

Drugs-of-Abuse Testing in Vitreous Humor

Barry S. Levine and Rebecca A. Jufer

Summary

Vitreous humor is a fluid contained in the eye that is largely composed of water. Its value for the postmortem analysis of ethanol has been well established. However, studies of drug disposition into vitreous humor are limited. This chapter reviews pertinent studies that have examined drug deposition into vitreous humor. The specific drugs/drug classes that are discussed include amphetamines, cocaine, THC, and opioids. Some of the advantages of vitreous humor as a matrix for drug analysis include the increased stability of certain drugs in this matrix and its amenability to analysis with little or no pretreatment. Disadvantages of vitreous humor analysis include a limited specimen volume and the limited interpretative value of analytical results.

Key Words: Vitreous humor, postmortem, toxicology.

1. STRUCTURE OF THE EYE

The human eye is recessed in the pyramid-shaped bony orbit and is connected to the brain by the optic nerve. The eyeball provides protection for the retina, the photosensitive portion of the eye. Both the retina and the optic nerve are enclosed by dense fibrous tissue, the sclera, and the dura mater. There are three chambers of the eye: the anterior chamber, the posterior chamber, and the vitreous. The dura mater surrounds the optic nerve and merges with the sclera to occupy the posterior chamber that constitutes five-sixths of the globe.

From: *Forensic Science and Medicine: Drug Testing in Alternate Biological Specimens*
Edited by: A. J. Jenkins © Humana Press, Totowa, NJ

The remaining one-sixth of the globe, the anterior portion, contains the cornea that refracts incident light. The vitreous is located between the lens and the retina and fills the center of the eye. It constitutes 80% of the eye and has a volume of approximately 4 mL *(1)*.

There are two fluids contained within the eye. The anterior chamber contains a clear, watery fluid, the aqueous humor. The aqueous humor is produced in the posterior chamber and has a volume of approximately 250 µL *(1)*. The vitreous is comprised of a transparent, delicate connective tissue gel called the gel vitreous or a transparent liquid called the liquid vitreous. The gel vitreous is a collagen gel that is water insoluble and liquefies with age such that the adult eye contains only liquid vitreous *(2)*. For the purposes of this chapter, the gel vitreous and the liquid vitreous will be considered as one specimen and will be referred to as the vitreous humor.

2. VITREOUS HUMOR COMPOSITION

The vitreous humor weighs approximately 4 g. It consists of 99% water and has a specific gravity of 1.0050–1.0089. It has a refractive index of 1.3341, which is slightly lower than the aqueous humor. Its viscosity is approximately two times that of water but with an osmotic pressure close to that of aqueous humor *(1)*. The pH of the vitreous humor is 7.5. The osmolality of the vitreous humor ranges from 288 to 323 mOsm/kg, slightly higher than the osmolality of serum (75–295 mOsm/kg) *(3)*.

Collagen is the major structural protein of the vitreous humor. Although similar to cartilage collagen, there are some distinct differences. Another major component of the vitreous humor is hyaluronic acid (HA). HA is a glycosamineglycan, a polysaccharide composed of repeating disaccharide units; each unit contains a hexosamine linked to uronic acid. The vitreous humor is composed of interpenetrating networks of HA molecules and collagen fibrils. In addition to collagen and HA, there are six specific non-collagenous proteins and two types of glycoproteins in the human vitreous humor *(3)*.

Besides the larger molecules, there are a number of low-molecular weight substances that are found in the vitreous humor. Vitreous humor concentrations of sodium and chloride will approximate the serum concentrations of these ions in healthy adults, especially in the early postmortem period. Potassium concentrations in the vitreous humor increase rapidly after death as potassium leaves the cells into nearby fluids. Vitreous humor calcium concentrations are comparable to serum calcium concentrations. The vitreous glucose concentration is similar to the serum concentration. Glucose is rapidly broken down after death in non-diabetics but remains elevated in uncontrolled diabetics. Two nitrogenous compounds, urea and creatinine, are also present in concentrations similar to serum and are stable during the early postmortem period. In postmortem cases, these substances are routinely measured as indicators for the serum concentrations of these substances at death *(4)*.

3. MOVEMENT OF SUBSTANCES INTO AND FROM VITREOUS HUMOR

There is a dual blood supply to the human eye. One is the retinal circulation that is mediated by the central artery and vein of the retina. The other source is the uveal circulation that is a system of ciliary arteries that run in the middle coat of the eye, the uvea *(5)*. The retinal capillaries are virtually impermeable to molecules with a molecular weight greater than 1900 Da. Movement across the blood–retinal barrier occurs primarily by diffusion. There also appears to be an active transport mechanism across the blood–retinal barrier *(2)*. The uveal circulation supplies the ciliary body and the iris. The ciliary body produces the vitreous humor. The ciliary capillaries have openings that permit macromolecules to escape. Conversely, the iris capillaries have much lower permeability *(5)*.

Equilibrium between blood and vitreous is slower than between the blood and other extracellular fluids. This suggests the presence of a barrier and is called the blood–vitreous barrier. This barrier consists of vascular endothelium and its basement membrane, stroma, and two layers of ciliary epithelium *(5)*. The movement of molecules in and out of the vitreous occurs by a number of mechanisms – diffusion, hydrostatic pressure, osmotic pressure, convection, and active transport. Water movement is significant, as approximately 50% of the water is replaced every 10–15 min. High-molecular weight substances and colloidal particles travel by convection. Low-molecular weight substances move in and out of the vitreous primarily by diffusion; however, there is evidence that bulk flow also contributes to their movement *(3)*.

Common drugs are small molecular weight molecules and therefore, move in and out of the vitreous by diffusion. Only free drug is able to leave the blood and enter the vitreous. Therefore, drugs that are not highly protein bound would be expected to have significant concentrations in the vitreous humor. Conversely, drugs that are highly protein bound would not appear to any significant extent in the vitreous humor.

4. SPECIMEN COLLECTION

The vitreous humor should be collected using a syringe and a 20-gauge needle. The needle is placed against the eye at the lateral aspect just above the junction between the upper and lower eyelids. The needle is inserted into the eye approximately 2 cm and the vitreous humor is gradually withdrawn. It is recommended that the vitreous humor from both eyes be collected. No preservatives need to be added to the specimen; however, it should be stored in the refrigerator until analyzed *(6)*.

5. DRUG ANALYSIS IN VITREOUS HUMOR

The analysis of drugs in vitreous humor is similar to the analysis in other postmortem fluids. Most of the published methods that were developed for blood and urine have been employed successfully with vitreous humor. These methods have used either solid-phase or liquid-liquid extraction depending on the characteristics of the analytes. Detection systems, such as gas chromatography, gas chromatography/mass spectrometry, liquid chromatography, and liquid chromatography/mass spectrometry have all been used to identify and quantify drugs in vitreous humor.

Because vitreous humor is approximately 99% water, some methods of analysis have been used following little or no sample preparation. For example, Chronister et al. *(7)* analyzed benzoylecgonine (BE) in vitreous humor by cloned enzyme donor immunoassay (CEDIA) without any specimen preparation. The manufacturer's urine calibrators were used, but a lower cutoff was programmed into the autoanalyzer. Felscher et al. used fluorescence polarization immunoassay for vitreous humor drug analysis without pretreatment *(8)*. Logan and Stafford tested vitreous humor for cocaine and BE by liquid chromatography by diluting the vitreous humor 1:1 with water and filtering through a 3 cm pre-column *(9)*.

6. CASE REPORTS AND INTERPRETATION OF RESULTS

6.1. Amphetamines and Hallucinogenic Amines

Limited data was identified in the scientific literature pertaining to the distribution of amphetamine and methamphetamine into vitreous humor. There were several single case reports of methylenedioxymethamphetamine (MDMA), where MDMA and/or methylenedioxyamphetamine (MDA) were measured in the vitreous humor. Crifasi and Long *(10)* reported MDMA and MDA concentrations in blood and vitreous humor in a traffic fatality. MDMA concentrations in the blood and vitreous humor were 2.14 and 1.11 mg/L, respectively; MDA concentrations were less than 0.25 mg/L in both specimens. Decaestecker et al. *(11)* measured MDMA concentrations in an overdose from 4-methylthioamphetamine. The subclavian blood, femoral blood, and vitreous humor MDMA concentrations were 0.013, 0.010, and 0.067 mg/L, respectively. DeLetter et al. *(12)* published an extensive fluid and tissue distribution study of

MDMA and MDA in an overdose from MDMA. Blood MDMA concentrations ranged from 3.1 to 7.6 mg/L depending on the site of the blood specimen. The vitreous humor MDMA concentration was 3.4 mg/L. Blood MDA concentrations ranged from 0.09 to 0.29 mg/L; the vitreous humor MDA concentration was 0.06 mg/L. In another case report from the same laboratory, blood MDMA and MDA concentrations ranged from 0.27 to 0.81 mg/L and 0.009 to 0.043 mg/L, respectively *(13)*. Vitreous humor MDMA and MDA concentrations were 0.36 and 0.015 mg/L, respectively. These studies indicate that MDMA and MDA were detectable in the vitreous humor when identified in the blood, but the data is too limited to draw any other conclusions.

6.2. Cannabinoids

As many postmortem laboratories do not test for delta-9-tetrahydrocannabinol (THC) or its metabolites, the data related to the detection of these compounds is limited. Lin and Lin *(14)* measured 11- nor-delta-9-tetrahydrocannabinol-9-carboxylic acid (THC-COOH) in blood and vitreous humor specimens from 50 driver fatalities. Alkaline hydrolysis was performed on all specimens. Vitreous humor concentrations of THC-COOH in all cases were less than 0.01 mg/L. Heart blood concentrations of THC-COOH ranged from 0.01 to 0.33 mg/L and exceeded the vitreous humor concentration in all cases. This is expected because the polarity of THC-COOH makes it difficult to cross the blood-vitreous barrier. THC was not tested in these cases. From this limited data, it appears that vitreous humor is not a useful specimen to screen for marijuana use.

6.3. Cocaine and Metabolites

There have been a number of studies published on the distribution of cocaine and metabolites in vitreous humor. Several of these studies describe single cases where blood and vitreous humor were tested for cocaine and/or BE. Sturner and Garriott *(15)* reported a single case of a cocaine overdose where the blood cocaine concentration was 8.5 mg/L and the vitreous humor cocaine concentration was 3.8 mg/L. BE concentrations were not reported. In another overdose case, Poklis et al. *(16)* found a blood cocaine concentration of 1.8 mg/L and a vitreous humor cocaine concentration of 2.4 mg/L. The cocaine was administered intravenously. Furnari *(17)* published a tissue distribution study in a cocaine intoxication death in a body packer. The blood and vitreous humor cocaine concentrations were 4.0 and 7.1 mg/L, respectively; the BE concentrations were 17.0 and 5.8 mg/L, respectively, in the blood and the vitreous humor.

In addition to these single case reports, a number of researchers have collected blood and vitreous humor concentrations in multiple cases. Logan and Stafford *(9)* used liquid chromatography to measure cocaine and BE in 28 paired blood and vitreous humor specimens. Cocaine concentrations greater than 0.01 mg/L were detected in 17 of the 28 blood specimens and 22 of the 28 vitreous humor specimens. All six specimens where the vitreous humor cocaine concentration was less than 0.01 mg/L were associated with a blood cocaine concentration less than 0.01 mg/L. The mean vitreous humor to blood cocaine concentration ratio was 1.61 with a range between 0.1 and 2.6. A comparison of the blood and vitreous humor cocaine concentrations showed considerable spread, with a correlation of $R = 0.70$. A high blood cocaine concentration was generally associated with a high vitreous humor cocaine concentration. BE was quantitated in only five of the 28 specimens, so a meaningful comparison to vitreous humor BE concentrations could not be performed. BE was tested in 24 of the 28 vitreous humor specimens at a limit of quantitation of 0.05 mg/L. The vitreous humor BE concentration exceeded the vitreous humor cocaine concentration in only three of these 24 specimens.

Mackey-Bojack et al. *(18)* quantitated cocaine, BE, and cocaethylene in blood and vitreous humor in 62 medical examiner cases. The average blood cocaine concentration in these cases was 0.489 mg/L with a range of 0–6.28 mg/L. This was not statistically different from the average vitreous humor cocaine concentration of 0.613 mg/L (range 0–4.50 mg/L). Differences between the average blood and vitreous humor cocaethylene concentrations were also found not to be statistically different. The mean blood cocaethylene concentration was 0.022 mg/L and the mean vitreous humor cocaethylene concentration was 0.027 mg/L. However, there was a statistically significant difference between the blood and vitreous humor BE concentrations; the average blood BE concentration was 1.941 mg/L and the average vitreous humor BE concentration was 0.989 mg/L. The blood BE concentrations exceeded the vitreous humor cocaine concentrations in 46 of the 53 cases. There were linear relationships between blood and vitreous humor concentrations for cocaine and BE but not for cocaethylene. Nevertheless, the authors of the study warned that prediction of a blood concentration from a vitreous humor concentration could not be reliably performed as the magnitude and direction of the concentration differences between the two specimens was too variable.

Chronister et al. *(7)* analyzed 392 vitreous humor specimens for BE by a commercially available immunoassay. Twenty-three specimens were positive at a cutoff of 0.1 mg/L. Twenty-two of these confirmed positive for BE by gas chromatography/mass spectrometry in the blood. The blood BE concentrations exceeded the vitreous humor concentrations in 19 of the 22 cases.

A number of conclusions can be made from the above discussion. Vitreous humor is a useful specimen to identify cocaine use. Either cocaine or BE can be used as the target compound to screen for cocaine use. Vitreous humor cocaine concentrations are generally in the same range as blood cocaine concentrations. However, vitreous humor BE concentrations are generally lower than blood BE concentrations. This is expected, as BE, being more hydrophilic than cocaine, would encounter greater difficulty in crossing the blood–vitreous barrier. It is generally unreliable to predict a blood cocaine or BE concentration from a vitreous humor cocaine or BE concentration.

6.4. Opioids

There have been a number of published studies that include data on the distribution of opioids into vitreous humor. The specific opioids for which data are available include morphine/heroin, oxycodone, methadone, propoxyphene, fentanyl, and sufentanil.

6.4.1. MORPHINE/HEROIN

Ziminski et al. *(19)* investigated 49 drug-related medical examiner cases, of which 13 were positive for morphine. Morphine was detected in vitreous humor from 12 of the 13 cases, ranging in concentration from 0.03 to 0.14 mg/L. The case in which morphine was not detected in vitreous humor was classified as an "acute opiate intoxication," and the authors suspected that death occurred prior to morphine equilibration with vitreous humor. There was also a single case in which morphine was detected in vitreous humor but not in blood. Of the 11 cases in which morphine was detected in both vitreous humor and blood, the vitreous humor to blood morphine concentration ratios ranged from 0.03 to 6.0, with a mean of 0.18 (median = 0.11).

A study of 10 cases of heroin overdose quantitated free and total morphine in vitreous humor by RIA *(20)*. Following treatment with β-glucuronidase, higher concentrations of morphine in vitreous humor were observed, demonstrating that morphine conjugates distribute into the vitreous humor. The ratio of free to total morphine ranged from 0.14 to 0.77. The authors also analyzed corresponding blood, urine and bile specimens and found that vitreous humor concentrations did not correlate well with the measured concentrations in these matrices. The vitreous humor to blood morphine concentration ratios for these 10 cases were all less than one, ranging from 0.08 to 0.83.

Gerostamoulos and Drummer *(21)* reported a series of 40 heroin-related deaths. The authors analyzed femoral blood, bile, brain, cerebrospinal fluid, kidney, liver, plasma, urine, and vitreous humor for free morphine (MOR), morphine-3-glucuronide (M3G), and morphine-6-glucuronide (M6G). The mean concentrations (mg/L) of morphine species in vitreous humor were M3G, 0.08; M6G, 0.02; free MOR, 0.09; and total MOR, 0.16. In addition, the authors calculated the percentage of M3G and Free MOR as a proportion of total morphine. Of all the specimens analyzed, vitreous humor contained the lowest proportion of M3G (32%) and the highest proportion of free morphine (70%). When comparing blood and vitreous humor-free morphine concentrations, a correlation coefficient of 0.261 was calculated, with a 95% confidence interval of –0.219–0.639, indicating that vitreous humor morphine concentrations may be of limited value for toxicological interpretation.

Pragst et al. *(22)* investigated the distribution of morphine in blood, vitreous humor, and cerebrospinal fluid in 154 opiate-related fatalities. In general, they found free and conjugated morphine concentrations to be lower in vitreous humor than in blood. Most frequently, vitreous humor to blood concentration ratios were in the range of 0.4 to 0.6 for free morphine and less than 0.3 for conjugated morphine. The authors also suggested that survival time significantly influenced vitreous humor to blood morphine concentration ratios. In cases for which a long (>5 h) survival time was known, they observed lower concentrations of morphine in vitreous humor and blood and greater vitreous humor to blood concentration ratios compared to the cases with a shorter survival time. However, the authors cautioned that the interpretative value of this finding is complicated by the effects of repeated use, which may result in the accumulation of morphine in the vitreous humor.

A smaller study completed by Bogusz *(23)* quantitated free morphine and M6G in four cases of heroin overdose. The author observed poor correlation between the free morphine to M6G ratios in blood and vitreous humor and no relationship between the estimated survival time and the free morphine to M6G ratio in either specimen type. However, the author indicated that any observations made with this study must be confirmed by a study including a greater number of cases. A second study by Bogusz et al. *(24)* analyzed vitreous humor, cerebrospinal fluid, and blood for morphine, M3G, and M6G in specimens collected from 21 heroin-related deaths; vitreous humor was available for 12 cases. It was observed that the M3G concentration was greater than the M6G concentration in all vitreous humor specimens analyzed. Vitreous humor M3G concentrations ranged from 36 to 250 ng/mL while M6G concentrations were between 3 and 46 ng/mL. It was also observed that vitreous humor concentrations of morphine, M3G, and M6G were lower than blood and cerebrospinal fluid concentrations of these analytes.

Lin et al. *(25)* analyzed vitreous humor from 223 opiate-positive cases to determine whether vitreous humor codeine and morphine concentrations could be useful for differentiating death due to codeine overdose from death due to heroin or morphine use. Heroin use was confirmed in 41 cases by the presence of 6-acetylmorphine (6-AM) in vitreous humor. All of these heroin-related cases had a vitreous humor-free codeine to free morphine ratio significantly less than 1. In contrast, 20 cases that were attributed to codeine overdose had a free codeine to free morphine ratio greater than or equal to 1. Morphine and codeine concentrations in blood were not reported. The authors concluded that the codeine–morphine distribution pattern in vitreous humor resembles that observed in blood and urine. When the codeine concentration in vitreous humor is sufficiently high, it can serve as a valid indicator for the differentiation of codeine- and morphine-induced fatalities. However, the authors noted that the generally low concentration of codeine in vitreous humor might limit the usefulness of these findings.

A recent study by Wyman and Bultman *(26)* analyzed specimens from 25 heroin-related deaths. 6-AM was detected in all vitreous humor specimens, and it was present at a greater concentration than blood or CSF in 21 of the 25 vitreous humor specimens. In addition, vitreous humor was the only specimen that was positive for 6-AM in nine cases. The authors concluded that the detection of 6-AM in postmortem cases could be greatly improved by the analysis of vitreous humor.

6.4.2. OXYCODONE

There have only been a few reports of oxycodone distribution in vitreous humor. Drummer et al. *(27)* published a study of nine oxycodone-related deaths. Vitreous humor was analyzed for one case, with a vitreous humor to blood oxycodone concentration ratio of 1.2 (B = 1.5 mg/L, VH = 1.8 mg/L).

Anderson et al. *(28)* have also completed a study of oxycodone-related deaths. Vitreous humor oxycodone concentrations were reported for seven cases. The mean vitreous humor to heart blood oxycodone concentration ratio for these cases was 1.1 (median = 1.0), with a range of 0.2–2.0. A ratio greater than or equal to 1 was observed in five of the seven cases. Vitreous humor oxycodone concentrations ranged from 0.18–0.82 mg/L.

In a study of 14 cases presented by Winterling et al. *(29)*, the mean vitreous humor to blood oxycodone concentration ratio was 1.7 (median = 1.7), with a range of 0.24–3.9. This ratio was ≥1.0 in eleven of the 14 cases. The vitreous humor oxycodone concentrations ranged from 29 to 900 ng/mL.

6.4.3. METHADONE

Methadone distribution into vitreous humor has been described in several studies. Sturner and Garriott *(15)* reported two cases of methadone intoxication. The blood methadone concentrations were 0.7 and 0.6 mg/L, whereas the vitreous humor methadone concentration was 0.1 mg/L in both cases. A study completed by Jennings *(30)* included the analysis of vitreous humor and blood specimens from 47 methadone-positive medical examiner cases. Methadone was the primary analyte detected in vitreous humor; EDDP was detected in only two of the 47 vitreous humor specimens. Vitreous humor methadone concentrations ranged from 32 to 840 ng/mL. Peripheral blood methadone concentrations ranged from 82 to 1800 ng/mL. The average vitreous humor to peripheral blood methadone concentration ratio was 0.29 (median = 0.25), with a range of 0.08 to 0.98. In general, the vitreous humor to blood methadone concentration ratio was less than 0.50, with only four of the 47 cases having a ratio exceeding 0.5.

6.4.4. PROPOXYPHENE

Limited data exists relating to the distribution of propoxyphene into vitreous humor. Sturner and Garriott *(15)* published data from 10 cases of propoxyphene intoxication. Propoxyphene was detected in vitreous humor from all cases in concentrations ranging from 0.5 to 5.0 mg/L. In these cases, the vitreous humor to blood propoxyphene concentration ratio ranged from 0.05 to 0.34. The authors suspected that low vitreous humor to blood propoxyphene concentration ratios were detected in some cases because death occurred prior to equilibrium between vitreous humor and blood.

6.4.5. FENTANYL

Anderson and Muto *(31)* published data from a series of 25 fentanyl-positive postmortem cases. Vitreous humor fentanyl concentrations were available for three of these cases and ranged from 8 to 20 ng/mL. Vitreous humor to blood fentanyl concentration ratios were greater than unity in each of these cases and ranged from 1.25 to 2.08.

6.4.6. SUFENTANIL

Ferslew et al. *(32)* published a single case report involving a subject who committed suicide by the intravenous administration of midazolam and sufentanil. In this case, the vitreous humor sufentanil concentration was 1.2 ng/mL and the blood concentration was 1.1 ng/mL, resulting in a vitreous humor to blood concentration ratio close to unity.

6.4.7. OPIOID SUMMARY

Opioid distribution into vitreous humor is variable and dependent on the protein binding characteristics and lipophilicity of the specific opioid detected. Opioids including morphine, 6-AM, codeine, oxycodone, methadone, propoxyphene, and fentanyl have been detected in vitreous humor and may serve as indicators of use. The limited number of case studies available indicate that methadone, morphine, and propoxyphene are generally detected in vitreous humor at concentrations considerably less than blood concentrations of these analytes, while vitreous humor oxycodone and fentanyl concentrations are closer to blood concentrations. In addition, vitreous humor has been shown to be a useful marker for heroin use when urine is unavailable for 6-AM confirmation. As vitreous humor is devoid of esterase activity, 6-AM is relatively stable in this matrix compared to blood. The interpretative value of opioid concentrations in vitreous humor is limited and are most valuable when they are available in addition to blood concentrations.

6.5. Phencyclidine

At this time, a literature search could not identify any published studies that describe the distribution of phencyclidine into vitreous humor.

7. ADVANTAGES AND DISADVANTAGES

There are a number of advantages to the analysis of drugs in vitreous humor. The specimen is easily collected during the postmortem examination; even if a complete autopsy is not performed vitreous humor can be obtained. The specimen is clear and serous and consists essentially of water. As a result, the specimen is easy to work with analytically. Any method developed for urine or more complex postmortem specimens should be amenable to vitreous humor analysis. Commercially available immunoassays developed for urine can be used on vitreous humor. For chromatographic procedures, less specimen preparation is required.

As vitreous humor is located in an anatomically isolated position, it is more protected from putrefaction, charring, and trauma than more centrally located fluids and tissues. During the putrefactive process, decomposition products such as tyramine and phenethylamine may interfere with both the extraction and analysis of many drugs in blood and tissue specimens. This is less of a factor with vitreous humor drug analysis. However, when the body decomposes, the vitreous becomes desiccated, making it more difficult to collect the specimen. Moreover, if death involves trauma to central organs, the only specimen that may be available is cavity blood, with its potential contamination from tissues and stomach contents. In these cases, vitreous humor may be a useful alternative specimen.

There are also some disadvantages in using vitreous humor for drug analysis. The main drawback is specimen volume. At best, only about 5 mL of vitreous humor can be collected on a case. If multiple drug analyses are required, it may be necessary to use less sample volume for each assay, limiting assay sensitivity. Some assays may not be performed at all because of limited sample volume. In addition, the need to perform electrolyte and glucose (Chem 7) analysis may over ride drug testing decisions.

Another disadvantage to the use of vitreous humor for drug analysis is a limited database from which interpretive assessments of the analytical results can be made. At the present time, the utility of vitreous humor in drug analysis is as an adjunct to blood or tissue drug analysis.

REFERENCES

1. Tripathi RC, Tripathi BJ. Anatomy of the human eye, orbit and adnexa. In: Davson H, editor. *The Eye*. Orlando, FL: Academic Press, 1984: 1–268.
2. Balazs EA, Denlinger JL. The vitreous. In: Davson H, editor. *The Eye*. Orlando, FL: Academic Press, 1984: 533–589.
3. Sebag J. *The Vitreous: Structure, Function and Pathobiology*. New York: Springer-Verlag, 1989.
4. Coe JI. Postmortem chemistry update. Emphasis on forensic application. *Am J Forensic Med Pathol* 1993; 14(2):91–117.
5. Cole DF. Ocular fluids. In: Davson H, editor. *The Eye*. Orlando, FL: Academic Press, 1984: 269–390.
6. Bost RO. Analytical toxicology of vitreous humor. In: Yong SHY, Sunshine I, editors. *Handbook of Analytical Therapeutic Drug Monitoring and Toxicology*. Boca Raton, FL: CRC Press, 1997: 281–302.
7. Chronister CW, Walrath JC, Goldberger BA. Rapid detection of benzoylecgonine in vitreous humor by enzyme immunoassay. *J Anal Toxicol* 2001; 25(7):621–624.
8. Felscher D, Gastmeier G, Dressler J. Screening of pharmaceuticals and drugs in synovial fluid of the knee joint and in vitreous humor by fluorescence polarization immunoassay (FPIA). *J Forensic Sci* 1998; 43(3):619–621.
9. Logan BK, Stafford DT. High-performance liquid chromatography with column switching for the determination of cocaine and benzoylecgonine concentrations in vitreous humor. *J Forensic Sci* 1990; 35(6):1303–1309.
10. Crifasi J, Long C. Traffic fatality related to the use of methylenedioxymethamphetamine. *J Forensic Sci* 1996; 41(6):1082–1084.
11. Decaestecker T, De Letter E, Clauwaert K, Bouche MP, Lambert W, Van Bocxlaer J et al. Fatal 4-MTA intoxication: development of a liquid chromatographic-tandem mass spectrometric assay for multiple matrices. *J Anal Toxicol* 2001; 25(8): 705–710.
12. De Letter EA, Clauwaert KM, Lambert WE, Van Bocxlaer JF, De Leenheer AP, Piette MH. Distribution study of 3,4-methylenedioxymethamphetamine and 3,4-methylenedioxyamphetamine in a fatal overdose. *J Anal Toxicol* 2002; 26(2): 113–118.
13. De Letter EA, Bouche MP, Van Bocxlaer JF, Lambert WE, Piette MH. Interpretation of a 3,4-methylenedioxymethamphetamine (MDMA) blood level: discussion by means of a distribution study in two fatalities. *Forensic Sci Int* 2004; 141 (2–3):85–90.
14. Lin DL, Lin RL. Distribution of 11-nor-9-carboxy-delta 9-tetrahydrocannabinol in traffic fatality cases. *J Anal Toxicol* 2005; 29(1):58–61.

15. Sturner WQ, Garriott JC. Comparative toxicology in vitreous humor and blood. *Forensic Sci* 1975; 6(1–2):31–39.
16. Poklis A, Mackell MA, Graham M. Disposition of cocaine in fatal poisoning in man. *J Anal Toxicol* 1985; 9(5):227–229.
17. Furnari C, Ottaviano V, Sacchetti G, Mancini M. A fatal case of cocaine poisoning in a body packer. *J Forensic Sci* 2002; 47(1):208–210.
18. Mackey-Bojack S, Kloss J, Apple F. Cocaine, cocaine metabolite, and ethanol concentrations in postmortem blood and vitreous humor. *J Anal Toxicol* 2000; 24(1):59–65.
19. Ziminski KR, Wemyss CT, Bidanset JH, Manning TJ, Lukas L. Comparative study of postmortem barbiturates, methadone, and morphine in vitreous humor, blood and tissue. *J Forensic Sci* 1984; 29(3):903–909.
20. Bermejo AM, Ramos I, Fernandez P, Lopez-Rivadulla M, Cruz A, Chiarotti M et al. Morphine determination by gas chromatography/mass spectroscopy in human vitreous humor and comparison with radioimmunoassay. *J Anal Toxicol* 1992; 16(6):372–374.
21. Gerostamoulos J, Drummer OH. Distribution of morphine species in postmortem tissues. Ferrara DS, editor. *Proceedings from the XXXV Annual Meeting of The International Association of Forensic Toxicologists*, 33–36, 1997.
22. Pragst F, Herre S, Scheffler S, Hager A, Leuschner U. Comparative investigation of drug concentrations in cerebrospinal fluid, vitreous humor and blood. Spiehler V, editor. *Proceedings of the TIAFT/SOFT Joint Congress on Forensic Toxicology*, 281–291, 1995, Newport Beach.
23. Bogusz MJ. Concentrations of morphine and its glucuronides among fatally poisoned heroin addicts and patients during oral morphine therapy. Spiehler V, editor. *Proceedings of the TIAFT/SOFT joint Congress on Forensic Toxicology*, 1–11, 1994, Newport Beach.
24. Bogusz MJ, Maier RD, Driessen S. Morphine, morphine-3-glucuronide, morphine-6-glucuronide,and 6-monoacetylmorphine determined by means of atmospheric pressure chemical ionization-mass spectrometry-liquid chromatography in body fluids of heroin victims. *J Anal Toxicol* 1997; 21(5):346–355.
25. Lin DL, Chen CY, Shaw KP, Havier R, Lin RL. Distribution of codeine, morphine, and 6-acetylmorphine in vitreous humor. *J Anal Toxicol* 1997; 21(4):258–261.
26. Wyman J, Bultman S. Postmortem distribution of heroin metabolites in femoral blood, liver, cerebrospinal fluid, and vitreous humor. *J Anal Toxicol* 2004; 28(4):260–263.
27. Drummer OH, Syrjanen ML, Phelan M, Cordner SM. A study of deaths involving oxycodone. *J Forensic Sci* 1994; 39(4):1069–1075.
28. Anderson DT, Fritz KL, Muto JJ. Oxycontin: the concept of a "ghost pill" and the postmortem tissue distribution of oxycodone in 36 cases. *J Anal Toxicol* 2002; 26(7):448–459.
29. Winterling JM, Callery RT, Caplan MC, Perlman AS, Tobin JG, Jufer RA. Oxycodone-related deaths in Delaware. *Proceedings of the American Academy of Forensic Sciences IX*, 307–308. 2003.
30. Jennings JA. Distribution of methadone and EDDP in postmortem toxicology cases. Michigan State University, 2005 [Dissertation].

31. Anderson DT, Muto JJ. Duragesic transdermal patch: postmortem tissue distribution of fentanyl in 25 cases. *J Anal Toxicol* 2000; 24(7):627–634.
32. Ferslew KE, Hagardorn AN, McCormick WF. Postmortem determination of the biological distribution of sufentanil and midazolam after an acute intoxication. *J Forensic Sci* 1989; 34(1):249–257.

Chapter 8

Drugs in Bone and Bone Marrow

Olaf H. Drummer

Summary

Bone and bone marrow are specimens recently investigated as a matrix for drug testing. Following extraction by soaking bone in organic solvent, routine drug assays may be utilized to measure compounds. Antidepressants, benzodiazepines, and illicit drugs such as cocaine have been reported in bone.

Key Words: Drugs, bone, bone marrow, skeletonized remains, teeth, analysis.

1. INTRODUCTION

Bone and bone marrow have received relatively little attention compared to other alternative specimens. This is most likely due to the restriction of obtaining this specimen in post-mortem cases. Nevertheless, a number of case reports and studies have been conducted that clearly show that drugs are present in both bone and bone marrow. The presence of drugs in reported cases have aided the investigation into these deaths that was not possible given the often skeletonized state of the remains.

This chapter reviews the current state of knowledge both in terms of the drugs detected in these specimens and how drugs are recovered from these most unusual matrices.

2. PHYSIOLOGY AND STRUCTURE

Bone marrow is a vascular tissue that is present in the central cavities of bones. Marrow is most easily obtained from the major bones, ribs, and

From: *Forensic Science and Medicine: Drug Testing in Alternate Biological Specimens*
Edited by: A. J. Jenkins © Humana Press, Totowa, NJ

vertebrae. There are two types of marrow: red and yellow. Red marrow supports clusters of hemapoietic cells, white blood cells, macrophages, and has a rich blood supply. Yellow marrow supports numerous blood vessels and fat cells.

Bone is a highly vascularized tissue consisting of porous mineralized structure consisting of hydroxyapatite. The structure and composition varies according to the location. Cortical bones have a low turn-over rate and represents about 80% of the overall skeletal mass and provide the strength of the skeleton. Trabecular or cancellous bone is less dense, is spongy, and has a higher turn-over rate and largely consists of epiphyseal and metaphyseal parts of long bones and within smaller bones.

The degree of contact of drugs to the bone structures depends on the anatomical location of the bone and the local blood supply. The long bones (i.e., femur) receive the most blood supply while the short, flat, or irregular bones receive more superficial supply through the periosteum. Bone is not a uniform structure. For example, bone consists of layers or bundles of bone. Bone contains channels that contain small blood vessels and nerves (haversian canals).

3. TREATMENT OF BONE AND BONE MARROW FOR ANALYSIS

Marrow when collected as a fluid can be diluted with water or a buffer, the mixture macerated, and drugs extracted with a solvent in a very similar manner to other fluid specimens *(1)*. Bone marrow has a high fat content that may cause some difficulties, but this may be overcome by treating the dried solvent extract with hexane/ethanol (7:2) (e.g., 5 mL) and a small volume of water (0.2 mL). The hexane layer is discarded and the drugs isolated from the ethanol fraction (Table 1) *(2)*.

Table 1
Extraction Techniques in Bone and Bone Marrow

Tissue	Preparation and extraction technique
Bone including teeth	Prolonged soaking of annular rings or fragments in methanol
	Crushing bone and solvent extraction
Bone marrow	Solvent extraction of diluted marrow
	Treatment of marrow or dried marrow with methanol

It is difficult to separate bone marrow from the bone itself in substantially decomposed or skeletonized remains when only dry material is present in the core of the bone. Indeed it may not be necessary to separate dried material in the core from the bone if evidence of exposure to drugs only is desired. Treatment of the crushed material with a solvent can be sufficient to isolate drugs particularly if contact time with the solvent is at least several hours (Table 1).

The extraction of drugs from bone is much more difficult given the hard matrix. Although any bone can potentially be used, the more useful bones appear to be the femur or other long bones, where sectioning of the bones can occur to increase the surface area. Other than the long bones, the vertebrae and iliac crest are also samples collected at autopsy. Bones should first be thoroughly cleaned of any attached tissue and any surface contamination before attempting to extract drugs from the internal structures. Then annular rings of femur can be soaked in methanol (or another solvent) over an extended period of time (overnight). This usually extracts much of the drug contained within the porous matrix. Alternatively, bone can be crushed and treated with solvent. This will probably increase the extractability of drugs but can introduce some occupational health and safety dangers for the analyst when crushing bone. As quantitative measurements are not likely to be meaningful, it is not necessary to overdo the preparation of bone. Extracts of marrow or bone can be subject to the usual screening techniques, such as immunoassay or chromatographic techniques, providing appropriate comparisons are made with known drug-free control cases.

4. DRUG DETECTION

A range of drugs have been detected in both marrow and bone. These are summarized in Table 2. Drugs-of-abuse such as a number of amphetamines, cocaine, morphine and a number of benzodiazepines have all been detected in both marrow and bony structures. Benzoylecgonine has also been detected in bone; however 6-acetylmorphine, the heroin metabolite has not been detected in bone of heroin users. Numerous antidepressants and other opioids have also been detected suggesting that there is no apparent restriction in the ability of these tissues to take up drugs from the circulation.

Studies that have examined both blood and bone collected from the same decedent have not always shown good concordance (1–3). That is, some drugs in blood were not found in bone and some drugs in bone were not found in blood taken at autopsy. This is not surprising as drugs absorbed into bone from previous exposures will almost certainly remain in bone longer than in blood.

Table 2
Drugs Detected in Bone and Bone Marrow

Tissue	Drugs Detected
Bone marrow	Ethanol, bromoisoval, diazepam, flurazepam, triazolam, amitriptyline, nortriptyline, desipramine, doxepin, sertraline, moclobemide, cocaine, methamphetamine, amphetamine, morphine, and paraquat
Bone (including skeletonized remains, and teeth)	Aminopyrine, cyclobarbital, morphine, cocaine, oxycodone, anti-depressants, chlorpromazine, benzodiazepines, and other hypnotics

The time frame for retention of drugs in bone is unknown but is likely to be quite long since the turnover of bone is relatively slow.

A number of studies have shown relatively good correlation of bone marrow concentration with circulating blood in animal experiments using controlled doses. Indeed, marrow concentrations were detected within minutes of exposure of the animals to drug. This confirms the good perfusion of bony structures with the circulating blood and should enable drugs to be deposited in bone relatively quickly after exposure.

However, there is little known about the extent of uptake of most drugs. Controlled animal experiments and human cases where paired bone and blood specimens have been available show some drugs that are present in much higher concentrations than blood. For example, desipramine, flurazepam, and pentobarbital are present in concentrations some 30 times > that of blood *(2,4,5)*.

5. SKELETONIZED REMAINS

A number of case reports of remains found years after death have yielded drugs when the bone or bone containing dried marrow was subject to extraction techniques (Table 3). For example, methamphetamine and amphetamine were detected after 5 years in dry marrow *(6)*, and triazolam was detected in a suspected homicide victim 4 years after death *(7)*.

The relative resistance of drugs in bones to be degraded or leached out by prolonged exposure is also confirmed by the detection of methamphetamine and amphetamine in marrow of bones deliberately stored under water for 1 year *(13)*. It is likely, however, that the benzodiazepines have limited stability

Table 3
Examples of drug detection in skeletonized remains

Reference	Drugs detection and circumstances
Bösche and Burger *(8)*	Bromide in skeltonized remains possibly from carbromyl, bromisoval, and acecarbromal
Noguchi et al. *(9)*	Amitriptyline in vertebral bones
Kojima et al. *(6)*	Methamphetamine and amphetamine after 5 years in dry marrow
Bal et al. *(10)*	Acetaminophen and propoxyphene in dry marrow after burial for 2 years
Wohlenberg et al. *(11)*	Nortriptyline in bone of suspected suicide
Kudo et al. *(7)*	Triazolam in suspected homicide 4 years after death
Maeda et al. *(12)*	Bromisoval detected in dry bone marrow of suspected overdose

in bone structures over prolonged periods of time (months) *(14)*. Raikos et al. *(15)* reported 54% loss of morphine (155 ng/g compared with 340 ng/g) in bone buried for 1 year in a suspected heroin intoxication case.

6. TEETH

As teeth are another form of bone, it is not surprising that drugs have been found in this tissue *(16)*. Solvent extraction of pulverized teeth of known drug users has resulted in the detection of 6-acetylmorphine and morphine, as well as cocaine and benzoylecgonine *(17)*. It is likely that other drugs consumed by the owner will also be deposited. However, the rate and degree of entry into the various structures of teeth (dentine and enamel) is not known and likely to be slower than for long bones due to the lesser degree of vascularization.

7. ADVANTAGES AND DISADVANTAGES

Bones of various type and size including teeth contain evidence of past drug use. Specimens are relatively easy to collect at autopsy and occasionally are the only specimens available. Preparation by soaking is easy although crushing or pulverization requires additional tools. Conventional isolation and chromatographic techniques demonstrate the necessary sensitivity to detect the presence of drugs. Drug concentrations in bone may not correlate with post-mortem blood concentrations. Indeed, some drugs such as olanzapine, may not be detected in bone *(18)*. Although it is not possible at the present time to estimate the time or duration of exposure, drug presence in bone structures may at least provide evidence of prior contact with these substances.

REFERENCES

1. McIntyre LM, King CV, Boratto M, Drummer OH. Post-mortem drug analyses in bone and bone marrow. *Ther Drug Monit* 2000;22(1):79–83.
2. Winek CL, Westwood SE, Wahba WW. Plasma versus bone marrow desipramine: a comparative study. *Forensic Sci Int* 1990;48(1):49–57.
3. McGrath KK, Jenkins AJ. Analysis of postmortem bone/bone marrow specimens for drugs of importance in forensic toxicology. In: *Proceedings of the Joint TIAFT/SOFT Annual Meeting*, Washington DC, 2004.
4. Winek CL, Costantino AG, Wahba WW, Collom WD. Blood versus bone marrow pentobarbital concentrations. *Forensic Sci Int* 1985;27(1):15–24.
5. Winek CL, Pluskota M, Wahba WW. Plasma versus bone marrow flurazepam concentration in rabbits. *Forensic Sci Int* 1982;19(2):155–163.
6. Kojima T, Okamoto I, Miyazaki T, Chikasue F, Yashiki M, Nakamura K. Detection of methamphetamine and amphetamine in a skeletonized body buried for 5 years. *Forensic Sci Int* 1986;31(2):93–102.
7. Kudo K, Sugie H, Syoui N, Kurihara K, Jitsufuchi N, Imamura T, et al. Detection of triazolam in skeletal remains buried for 4 years. *Int J Legal Med* 1997;110(5): 281–283.
8. Bosche J, Burger E. Bromide-content in bones after poisoning with sedatives of bromine content. *Beitr Gerichtl Med* 1974;32:185–186.
9. Noguchi TT, Nakamura GR, Griesemer EC. Drug analyses of skeletonizing remains. *J Forensic Sci* 1978;23(3):490–492.
10. Bal T, Hewitt R, Hiscutt A, Johnson B. Analysis of bone marrow and decomposed body tissue for the presence of paracetamol and dextropropoxyphene. *J Forensic Sci Soc* 1989;29:219–223.
11. Wohlenberg N, Lindsey T, Backer R, Nolte K. Nortriptyline in maggots, muscle, hair, skin and bone in skeletonized remains. *TIAFT Bulletin* 1992;22(3):19–22.
12. Maeda H, Oritani S, Nagai K, Tanaka T, Tanaka N. Detection of bromisovalum from the bone marrow of skeletonized human remains: a case report with a comparison between gas chromatography/mass spectrometry (GC/MS) and high-performance liquid chromatography/mass spectrometry (LC/MS). *Med Sci Law* 1997;37(3):248–253.
13. Nagata T, Kimura K, Hara K, Kudo K. Methamphetamine and amphetamine concentrations in postmortem rabbit tissues. *Forensic Sci Int* 1990;48(1):39–47.
14. Gorczynski LY, Melbye FJ. Detection of benzodiazepines in different tissues, including bone, using a quantitative ELISA assay. *J Forensic Sci* 2001;46(4): 916–918.
15. Raikos N, Tsoukali H, Njau SN. Determination of opiates in postmortem bone and bone marrow. *Forensic Sci Int* 2001;123(2–3):140–141.
16. Cattaneo C, Gigli F, Lodi F, Grandi M. The detection of morphine and codeine in human teeth: an aid in the identification and study of human skeletal remains. *J Forensic Odontostomatol* 2003;21(1):1–5.

17. Pellegrini M, Casa A, Marchei E, Pacifici R, Mayne R, Barbero V, et al. Development and validation of a gas chromatography-mass spectrometry assay for opiates and cocaine in human teeth. *J Pharm Biomed Anal* 2005;40(3):662–668.
18. Horak, EL, Jenkins AJ. Postmortem tissue distribution of olanzapine and citalopram in a drug intoxication. *J Forensic Sci* 2005;50(3):679–681.

Chapter 9

Drugs-of-Abuse in Liver

Graham R. Jones and Peter P. Singer

Summary

The liver is the largest organ in the human body and has been used extensively as an important specimen in postmortem toxicology analysis. This chapter describes the advantages and disadvantages of using liver as a specimen for the detection and measurement of drugs-of-abuse in postmortem cases. The liver comprises relatively soft tissue amenable to the preparation of homogenates but contains high concentrations of lipids that may interfere in some analytical procedures. As a specimen, liver has the advantage that it is relatively unaffected by postmortem redistribution compared with blood, but drug concentrations in the lobe proximal to the stomach may be affected by postmortem diffusion in cases of oral overdose. The biggest impediment to the routine use of liver for the interpretation of positive drug findings is the lack of a comprehensive database of liver concentrations. The data contained here may assist those wishing to interpret liver concentrations of drugs-of-abuse. As for all drugs and specimens, the process of interpretation should include consideration of all aspects of the death investigation, including, as necessary, analysis of multiple specimens.

Key Words: Postmortem, liver, drugs-of-abuse.

1. INTRODUCTION

By definition, this chapter deals with the use of liver for the postmortem determination of drug exposure or intoxication prior to death. The liver is the largest organ in the body (other than the skin or muscle) and is easy to macerate. It has been used for the isolation and identification of poisons by forensic

From: *Forensic Science and Medicine: Drug Testing in Alternate Biological Specimens*
Edited by: A. J. Jenkins © Humana Press, Totowa, NJ

toxicologists for well over a century. Many drugs, especially those with chemically "basic" character (e.g., alkaloids), tend to concentrate in the liver 10- to 100-fold greater than in the blood, making it easier to isolate extracts that could be used in basic pharmacological tests, or if enough substance was present, to isolate and characterize crystalline material. Of more relevance today, liver tissue can serve a useful alternate specimen if blood is not available in some types of cases (decomposition, severe fire, and exsanguination). However, another important reason that liver is still widely used as a second or third specimen after blood is that postmortem blood drug concentrations are seldom static or homogeneous after death. For many drugs, concentrations in blood can easily increase twofold and often 10-fold or more over perimortem concentrations due to postmortem redistribution or postmortem diffusion. Therefore, measuring liver drug concentrations provides additional data to assist with interpretation.

Drugs-of-abuse that are chemically "basic" may be easily analyzed in the liver, including the amphetamines, ketamine, phencyclidine (PCP), and non-amphoteric analgesics such as methadone, meperidine, codeine, hydrocodone, and oxycodone. The so-called back-extraction procedures work well for basic drugs that have a relatively high pKa and can provide extracts that are virtually free of endogenous compounds such as lipids or fatty acids.

Chemically neutral compounds such as delta-9-tetrahydrocannabinol (THC) are considerably more challenging to accurately quantify in the liver because they are present at low concentrations and difficult to separate from other ionically neutral endogenous components. Morphine and hydromorphone, both amphoteric narcotic analgesics, are typically not analyzed for in liver because it is more difficult to obtain clean extracts, and as for most drugs, the concentrations measured are more difficult to interpret than in blood.

2. THE LIVER

The liver is a large, encapsulated, highly vascular organ situated predominately in the upper right front portion of the abdomen that typically weighs around 1.5 kg. It is divided into two lobes; the smaller left lobe partially overlays the stomach in the central to left portion of the abdomen; the right lobe occupies the upper right abdomen. The liver performs several critical functions, including the metabolism of food, the formation and storage of glycogen, and release, when required, as glucose. The other primary function of the liver is the metabolism of endogenous and exogenous toxicants. The liver is the primary site for drug metabolism in the body and contains the largest quantity of the critical cytochrome enzyme system, as well as liver alcohol dehydrogenase and many other enzymes.

From the perspective of drug absorption and metabolism, the hepatic portal system is the most important. All substances absorbed into the body through the digestive tract ultimately pass through the liver through the portal vein. The portal vein is fed by mesenteric and other venous systems from the stomach, the large and small intestines, the pancreas, and the spleen. Virtually, all blood returning to the heart and lungs through the venous system passes through the liver through the inferior vena cava; the liver is infused with arterial blood through the descending aorta.

3. KINETICS OF DRUG UPTAKE AND DISTRIBUTION

For decades, toxicologists have sought to find a relationship between blood and liver concentrations. The assumption has often been that the ratio between drug concentrations in the blood and liver approximates to some particular value, dependent largely on the physical characteristics and pharmacokinetics of the drug. However, interpretation of drug concentrations in liver, or for that matter any other organ, is far from straightforward. Consideration of the kinetics involved shows why a blood : liver ratio can vary wildly, not just because of postmortem changes in the blood concentration but due to the kinetics of the drug in the body, especially as it relates to absorption, distribution, and excretion.

For example, it is a fundamental principle of pharmacokinetics that the liver concentration of a drug during the absorption phase will vary significantly depending on whether it is administered orally or systemically (e.g., intravenous or intramuscular). With oral administration, the drug passes into the stomach where some absorption may occur, but for the majority of drugs, most absorption occurs in the small intestine. Thus, virtually all of an orally ingested drug passes through the mesentery vessels of the small intestine through the portal vein and into the liver. However, with intravenous or intramuscular administration, all of the drug initially enters the systemic circulation directly, and only after it has passed through a large portion of tissue in the body, does the remainder of it pass through the liver. Therefore, it is to be expected that the liver concentration of virtually any drug will vary at the time of peak blood concentration depending on the route of administration and dose. In other words, peak liver concentrations of a drug can be expected to be significantly higher after oral administration than after systemic administration during the period of initial distribution. In relation to the blood concentration of a drug, this gives rise to the "first pass effect" where a proportion of the drug never reaches the general circulation due to substantial metabolism in the liver.

In addition to these simple pharmacokinetic principles, it has to be considered that in the case of oral ingestion of a drug, postmortem diffusion through the stomach wall to the overlaid left lobe of the liver will increase the ratio of liver to blood concentration, especially if the mass of drug in stomach is relatively large (1,2). As far back as 1971, it was shown that in a rat model, concentrations of barbiturates in the liver increase with time after death due to diffusion from the stomach (3). This has been demonstrated more recently for amitriptyline, also in a rat model (4). The same authors have shown that limited redistribution of drug from the lungs to the liver may occur postmortem although the magnitude of this increase is not great compared to that for blood. In other words, the drug concentration in the liver is already high compared to that in the pre-mortem blood and therefore, the magnitude of increase of the liver concentration will not be as great as it is in the blood (5). Blood is non-homogeneous, with drug concentrations frequently site dependent and the concentration of a drug in the liver may vary considerably depending on the specific portion of the liver that is sampled (e.g., which lobe and even which part of that lobe). While it has been repeatedly advocated that the site of collection of a blood sample should be documented, on only a few occasions has it been advocated that the exact site of collection of a liver sample be stated.

Therefore, the concentration of a drug in the liver after oral ingestion will depend on the dose ingested, the time after ingestion, the rate of passage of the drug through the stomach into the small intestine (including the time for stomach emptying), and the postmortem interval.

One additional consideration is to what extent a drug may continue to be metabolized after death. For example, it is known that cocaine is unstable in unpreserved postmortem specimens due to continued hydrolysis by stable esterases. However, much less is known regarding the extent to which other enzyme systems may continue to function in the early postmortem period.

When the variability of drug concentrations in blood is considered (e.g., femoral blood vs. cardiac or pulmonary blood), it is not surprising that for many drugs the ratio of concentration in blood versus liver varies considerably. Figure 1 shows the degree of variability that can occur in the ratio of codeine in blood versus liver in a large series of cases where death was attributed to various causes. Although this data includes blood from all sources, the correlation does not improve significantly when blood from only a femoral source is plotted.

3.1. Long-Term Drug Sequestration

One aspect of toxicology that has received little attention, and indeed is very difficult to study, is the long-term sequestration of drugs in the liver and other tissues. While this issue is not relevant to most death investigation cases, it is relevant to those where sensitive methodologies may detect small amounts of drugs and where those findings may affect the ultimate disposition of a

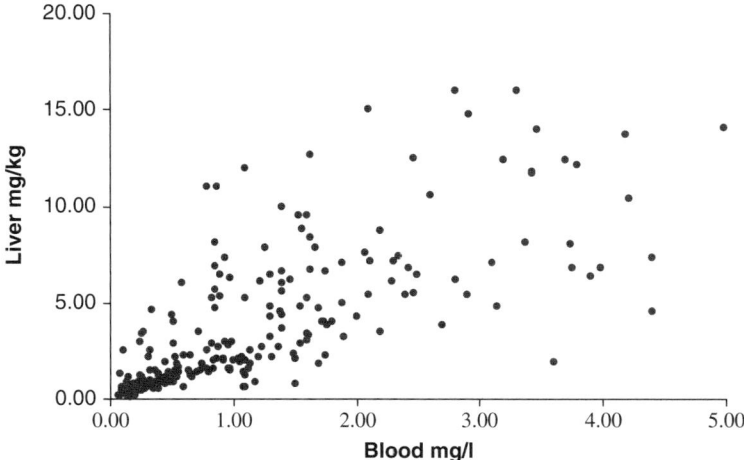

Fig. 1. Plot of codeine concentrations in liver versus corresponding postmortem blood for a random series of 236 medical examiner cases.

civil case. It is well documented that drug concentrations in the liver are often 50–100 times higher than concentrations in blood. It is also well known that the true terminal half-life of many drugs in the body is considerably longer than the traditional half-life in circulating blood. There is considerable data regarding the persistence of highly lipid-soluble substances such as chlorinated pesticides, and by extrapolation lipid-soluble drugs such as THC, although hard data for most drugs is anecdotal at best. However, given the lipid-soluble nature of many drugs and the high lipid content of the liver, the persistence of many drugs in the liver is likely to be considerably longer than in blood, and this should be taken into account when interpreting very low concentrations of drugs in the liver.

4. ANALYSIS OF DRUGS IN THE LIVER

Methods for the detection and quantitation of drugs in liver are not very different than for whole blood once the tissue is homogenized. Traditionally, homogenization of tissue is performed in a blender or homogenizer. Commercial units such as a probe-style homogenizer that combines shearing and ultrasonic action are preferred over food-processor style Waring blenders, because they are considerably more efficient in breaking down cell structures. Liver homogenates may be prepared in water or buffer at dilution from 1:1 to 1:10. If insufficient water or buffer is added, homogenization may be less efficient and the sample may be more difficult to pipette. Over-dilution may cause the concentration of the analyte to drop below the detection capability of the method. Homogenates may be centrifuged in an attempt to remove potentially interfering substances. This is not a problem for qualitative screening tests. However, if this is done for quantitative chromatographic tests, the internal standard must be added and allowed to equilibrate with the homogenate before centrifugation occurs. Care must also be taken to prevent cross-contamination of liver homogenates with homogenizing devices that can be difficult to thoroughly clean.

Most types of immunoassay can be used to screen liver homogenates for drugs-of-abuse although some may be subject to more interference than others. It is therefore important that such assays be investigated for specificity when using liver homogenates and that preferably matrix-matched controls be used.

Protease enzymes have been used in the past to liquefy liver homogenates *(6,7)*. However, this approach is seldom used today because it can increase the amount of extractable endogenous material and may not offer significant advantages over conventional liquid-liquid or solid-phase extraction (SPE) methods that use appropriately chosen internal standards.

A major consideration is whether an extraction and analytical procedure is selective enough to allow accurate detection and measurement of the target analyte without interference from the myriad of potentially interfering substances such as lipids or putrefactive amines. A procedure for basic drugs that incorporates a back-extraction will usually produce less interference and therefore more reliable results than one using a single-step extraction.

It also follows that the more decomposed the liver specimen, the more potential there is for interference. Therefore, the more specific the detection method, the more accurate the results are likely to be. However, one caveat to that statement is that non-specific interference that can occur with LC/MS due to ion suppression effects *(8–10)*.

The majority of drugs-of-abuse, including the amphetamines, cocaine, and most opiates (except morphine and hydromorphone), are moderately strong bases and may be isolated from liver homogenates by extraction into an organic solvent at basic pH and subsequent back-extraction into acid. In a typical procedure, basic drugs can be extracted into a suitable organic solvent after making the homogenate strongly basic (e.g., pH 11–12). The organic extract is transferred to another tube and strong acid added (e.g., 0.1–1 N sulfuric acid or hydrochloric acid). The aqueous phase containing the back-extracted basic drugs is transferred to another tube and basified with strong alkali (e.g., 2 N sodium hydroxide) and the drug extracted into a suitable organic solvent. This process will remove much of the neutral and acidic interference produced by fatty acids, lipids, and cholesterol. When drying the extracts, care must be taken to protect against loss of the volatile drugs such as amphetamines; some laboratories add methanolic hydrochloric acid to minimize such losses.

Morphine and hydromorphone are amphoteric and require a different approach requiring extraction at relatively mild alkaline pH (e.g., 8.0–9.0). Alternatively SPE methods may be used.

Kudo et al. *(11)* have published a method for the determination of free and total morphine in liver and other tissue using acid hydrolysis, extraction with an SPE (Extrelut NT) column, using dihydrocodeine as internal standard, followed by trimethylsilylation. The derivatized extract was submitted to GC/MS analysis in EI-SIM mode. The limit of detection of morphine was 0.005 mg/kg *(11)*. Cingolani et al. have demonstrated the successful extraction of morphine from formalin-preserved tissue. The mean levels of recovery of morphine in fixed tissues were 36.29% in liver and 74.93% in the formalin from liver *(12)*.

The major cocaine metabolite, benzoylecgonine, is relatively water soluble and not generally amenable to liquid-liquid extraction. However, benzoylecgonine and related metabolites may be extracted from liver and other tissues using SPE methods. Centrifugation of the liver homogenates after addition of an internal standard will facilitate passage through the SPE column.

5. INTERPRETATION

Interpretation of liver drug concentrations is difficult. The kinetics of drug absorption and distribution are such that the relationship between blood and liver concentrations is rarely consistent enough to allow reliable quantitative prediction of a single blood concentration from a liver concentration. The problem is compounded because well-defined reference ranges do not exist and much data for liver concentrations of drugs is typically dotted throughout the literature in a few individual case reports that maybe difficult to locate. One well-known book of drug monographs contains numerous references to liver concentrations of drugs *(13)*. However, those case reports are frequently overdose cases and relatively few are cases where it is known that "therapeutic" (or "recreational") doses of the drug are involved. The situation is further complicated because careful review of the original case reports is required to assess the contribution of other drugs, alcohol, or natural disease in the death.

In the absence of a reliable database in the literature, interpretation falls to individual laboratories that routinely measure liver drug concentrations in casework – for all types of death. A liver drug concentration can provide useful forensic information when considered with the available history and circumstances of the death, and preferably a blood concentration. For example, a higher than normal liver concentration can corroborate a high postmortem blood concentration and becomes even more useful when interpreted in conjunction with the case history. As with the interpretation of postmortem blood drug concentrations, it is critical that consideration should be given to the possibility of drug accumulation due to natural disease or impaired metabolism due to drug interactions or pharmacogenetic factors.

Following is a brief summary of the readily available literature regarding concentrations of some drugs-of-abuse in liver. For drugs where data on liver concentrations is sparse in the literature, data are included from the Alberta Medical Examiner's Office in Canada. However, it should be noted that no attempt has been made to differentiate cases where the presence of the drug is the cause of death or merely an incidental finding. Furthermore, the relationship between blood and liver concentrations is not necessarily linear, and indeed, there is a considerable amount of scatter when plotting blood versus liver concentrations for any drug (as seen in Fig. 1).

5.1. Amphetamines

5.1.1. AMPHETAMINE

There are several reports of amphetamine being measured in liver. In one review, a series of 11 amphetamine fatalities were reported with liver concentrations ranging from 4.3 to 74 mg/kg (average 30) with the corresponding blood concentrations 0.5 to 41 (average 8.6) mg/L *(14–20)*.

A distribution study of several amphetamines in a single fatality reported amphetamine concentrations in liver and femoral blood of 0.857 mg/kg and 0.198 mg/L, respectively *(21)*.

5.1.2. METHAMPHETAMINE

A fatal overdose ingestion of a large amount of methamphetamine resulted in a liver concentration of 206 mg/kg and blood concentration of 40 mg/L *(22)*. In another methamphetamine-related fatality, concentrations were 4.8 mg/kg in the liver and 2.0 mg/L in the blood *(23)*. Kojima reported two separate methamphetamine overdose cases with liver concentrations of 174 and 14.1 mg/kg with corresponding blood concentrations of 43 and 8.3 mg/L, respectively *(24,25)*. Moore reported the death of a 37-year-old man with postmortem methamphetamine concentrations of 2.2 mg/kg in liver and 0.68 mg/L in blood *(26)*. Sato reported methamphetamine concentrations in a burned body of 11.7 mg/kg in liver and 2.5 mg/L in blood *(27)*. Logan reported a massive overdose in a man swallowing a baggie of methamphetamine that produced liver and peripheral blood concentrations of 90.9 mg/kg and 53.7 mg/L, respectively *(28)*.

Bailey reported a large series of deaths involving methamphetamine, with and without concurrent cocaine use *(29)*. Methamphetamine liver and blood concentrations were 0.13–35.96 mg/kg (average 4.06) and 0.02–3.05 mg/L (average 0.90), respectively, without cocaine and 0–17.1 mg/kg (average 2.83) and 0–7.10 mg/L (average 1.03) with cocaine.

5.1.3. METHYLENEDIOXYAMPHETAMINE

Most MDA occurrences today tend to be as the metabolite of methylene-dioxymethamphetamine (MDMA) or co-ingested as an ingredient in illicit tablets. However, in the older literature, methylenedioxyamphetamine (MDA) was more common as a sole intoxicant. In one report of 12 cases, liver concentrations were 8–17 mg/kg (average 12) and the corresponding blood concentrations were 1.8–26 mg/L (average 9.3) *(30–36)*.

5.1.4. METHYLENEDIOXYMETHAMPHETAMINE

There have been at least three different cases where liver concentrations of MDMA have been reported. DeLetter reported an MDMA overdose with MDMA in liver at 26 mg/kg and in peripheral blood at 3.1 mg/L *(37)*. Sticht reported concentrations of 29.7 mg/kg in liver and 7.2 mg/L in peripheral blood in an overdose *(38)*. Dams found MDMA concentrations of 6.66 mg/kg in liver and 1.92 mg/L in femoral blood in a combined MDMA/*P*-methoxyamphetamine (PMA) fatality *(21)*.

5.1.5. P-METHOXYAMPHETAMINE

PMA is a fairly toxic amphetamine compound occasionally detected in forensic cases. Cimbura reported nine cases with liver concentrations ranging from 0.5 to 10 mg/kg and blood 0.3 to 1.9 mg/L *(39)*. Felgate reported a series of 10 PMA-related deaths with paired liver (femoral blood) concentrations of 11 (1.7), 1.4 (0.24), 7.4 (1.3), 7.1 (3.7), 21 (4.9), 5.6 (2.2), 6 (1.7), 7.5 (2.2), 6.0 (2.0), and 2.7 (0.53); average 7.8 mg/kg (2.0 mg/L in blood) *(40)*. PMA concentrations have been reported in three other fatalities with liver concentrations of 8.9, 6.8 and 18 mg/kg compared to 1.63, 0.4, and 1.8 mg/L in blood *(13,21,41)*.

5.2. Opiates and Opioids

5.2.1. MORPHINE AND HEROIN

Felby reported a series of ten fatalities following intravenous use of morphine only; liver total morphine concentrations were 0.4–18 mg/kg (average 3.0) and the corresponding blood concentrations 0.2–2.8 (average 0.7) *(42)*.

In a separate case, a man who died of trauma after 2 days in hospital and who received multiple morphine doses had a liver concentration of 0.11 mg/kg and blood concentration of 0.67 mg/L *(43)*. Cravey also reported blood and liver concentrations of morphine of 8.0 mg/L and 6.0 mg/kg, respectively, in a hospitalized patient *(44)*. However, it is not clear in these cases whether the morphine reported was unconjugated or total.

Chan reported two related morphine fatalities. One case involved oral ingestion with blood concentrations of 0.35 (free) and 1.53 mg/L total morphine and a liver total morphine of 7.0 mg/kg. The second case was due to intravenous administration and resulted in blood concentrations of 0.07 and 0.42 mg/L of free and total morphine, respectively, and 2.9 mg/kg total morphine in the liver *(45)*.

Moriya determined the tissue distribution of free and conjugated morphine in a man after injection with heroin and methamphetamine. Blood concentrations of free and total morphine varied from 0.462 to 1.350 and 0.534 to 1.570 mg/L depending on the site. A total morphine concentration of 4.20 mg/kg was found in the liver. 6-Acetylmorphine was not measured in the liver. In a separate in vitro experiment, the same authors found free and conjugated morphine to be stable in the blood and urine at 4, 22, and 37 °C for the 10-day study period. In the liver, conjugated morphine was converted almost completely to free morphine at 18–22 and 37 °C by the end of the 10 days although it was stable at 4 °C *(46)*.

Spiehler analyzed blood and liver for unconjugated and total morphine in 56 cases where the deceased died within 3 h of heroin/morphine use. The unconjugated morphine in blood ranged from 0.08 to 1.65 mg/L, total morphine from 0.10 to 2.2 mg/L, and liver total morphine from 0.41 to 5.5 mg/kg *(47)*. Using expert systems to analyze the data, Spiehler found blood unconjugated morphine, blood total morphine, and liver total morphine concentrations most useful for interpretation. Morphine overdoses were characterized by a blood unconjugated morphine greater than 0.24 mg/L and liver morphine concentration greater than 0.50 to 0.75 mg/kg *(48)*.

Kerrigan reported an unusual case involving a morphine pump and acquired tolerance. A 44-year-old male was receiving morphine for pain control. Quantitative analysis of free and total morphine determined unconjugated morphine concentrations in heart blood and liver of 96 mg/L and 88 mg/kg while total morphine concentrations were 421 mg/L and 256 mg/kg, respectively. Records indicate that the infusion pump may have continued to deliver the drug for 15–45 min following death. The cause of death was determined to be complication of adenocarcinoma of the pancreas, and the manner was natural *(49)*.

Lewis reported on the simultaneous analysis of multiple opiates in postmortem tissue, after civil aviation accidents, using SPE, TMS, oxime-TMS derivatives, and GC-MSD. Two cases contained morphine with total blood concentrations of 0.260 and 0.023 mg/L and liver concentrations of 0.87 and 0.083 mg/kg, respectively *(50)*.

5.2.2. CODEINE

Nakamura measured a large series of blood and liver concentrations. In the higher codeine concentrations (11 cases), the liver ranged from 0.6 to 45 mg/kg (average 6.8) with blood concentrations of 1.0 to 8.8 mg/L (average 2.8) *(51,52)*.

A series of 269 deaths associated with codeine in the Alberta Medical Examiner's system in the 6 years from March 1999 to September 2005 were analyzed. The liver concentrations ranged from 0.1 to 96 mg/kg (average 8.0, median 3.0, SD = 13.9), and the corresponding blood concentrations were 0.1 to 83 mg/L (average 2.6, median 1.1, SD = 6.5).

5.2.3. OXYCODONE

Sedgwick reported on an overdose in a diver with a liver concentration of 12 mg/kg and blood 5 mg/L *(53)*. Cravey reported two suicidal overdoses with liver concentrations of 22 and 63 mg/kg and corresponding blood concentrations of 4.3 and 14 mg/L *(54)*.

There are two reports of child deaths related to oxycodone. Armstrong et al. reported a 2-year-old girl with oxycodone as the only finding where the heart blood concentration was 1.36 mg/l and liver 0.2 mg/kg *(55)*. Levine reported the death of a 10-month-old baby boy with only oxycodone present. The blood and liver concentrations were 0.6 mg/L and 1.6 mg/kg, respectively *(56)*.

Lewis et al. reported an oxycodone case after a civil aviation accident with a blood concentration of 0.232 mg/L and a liver concentration of 0.755 mg/kg *(50)*.

A series of 126 deaths associated with oxycodone in the Alberta Medical Examiner's system in the 5 years from March 1999 to December 2004 were analyzed. The liver concentrations ranged from 0.12 to 480 mg/kg (average 10.0, median 1.4, SD = 48.7) and the corresponding blood concentrations were 0.02 to 520 mg/L (average 5.4, median 0.61, SD = 46.32).

5.2.4. HYDROCODONE AND HYDROMORPHONE

Lewis reported hydrocodone in four aircraft accident victims at liver concentrations of 0.142, 0.02, 0.447, and 0.082 mg/kg and corresponding blood concentrations of 0.022, 0.018, 0.102, and 0.036 mg/L, respectively *(50)*.

Baselt summarized data from eight fatal hydromorphone overdose cases with liver concentrations averaging 1.4 mg/kg (range 0.07–7.7) and blood concentrations averaging 0.3 mg/L (range 0.02–1.2) although data from at least two of the references were not readily available *(13,57,58).*

5.2.5. METHADONE

In a series of 10 deaths attributed to methadone, Manning et al report liver concentrations of 1.8–7.5 mg/kg (average 3.8) with corresponding blood concentrations of 0.4–1.8 mg/L (average 1.0) *(59).*

A series of 36 deaths due primarily to methadone in the Alberta Medical Examiner's system in the 5 years from June 1999 to August 2004 were analyzed. The liver concentrations ranged from 1.0 to 47 mg/kg (average 6.9, median 4.85, SD = 8.2) and the corresponding blood concentrations were 0.18 to 7.62 mg/L (average 1.0, median 0.505, SD = 1.3).

5.2.6. MEPERIDINE (PETHIDINE)

Siek reported the distribution of meperidine and its metabolite, normeperidine, in three oral and three intravenous deaths due to meperidine. For the oral cases, concentrations of meperidine and normeperidine in liver averaged 7 and 31 mg/kg (range 5–10 and 11–66) respectively; for the intravenous cases, liver concentrations average 8.3 and 7.3 mg/kg for (range 2–16 and 0–12) meperidine, and normeperidine respectively. The corresponding blood concentrations averaged 12 and 19 mg/L (range 8–20 and 8–30) for the oral route; blood concentrations average 4.3 and 2.5 mg/L (range 1–8 and 0–7) for the intravenous route *(60).*

A series of 33 deaths associated with meperidine in the Alberta Medical Examiner's system in the 7 years from March 1999 to March 2006 were analyzed. For 32 paired results, the liver concentrations ranged from 0.23 to 320 mg/kg (average 25.4, median 3.25, SD = 64) and the corresponding blood concentrations were 0.05 to 36.2 mg/L (average 3.2, median 1.2, SD = 6.5). For normeperidine, an active and potentially more toxic metabolite, liver concentrations ranged from 0.26 to 159 mg/kg (average 18.8, median 6.4, SD = 32) and the corresponding blood concentrations were 0.05 to 33.6 mg/L (average 3.3, median 0.72, SD = 6.4).

5.3. Phencyclidine and Ketamine

Baselt summarized 17 PCP cases where the liver concentrations averaged 23 mg/kg (range 0.9–170) and blood 4.8 mg/L (range 0.3–25) *(13,61–64).*

Peyton reported two ketamine fatalities with liver concentrations of 6.3 and 0.8 mg/kg and blood concentrations of 7.0 and 3.0 mg/L, respectively *(65).*

Moore reported a case of a mixed drug fatality involving ketamine and ethanol. The liver and blood ketamine concentrations were 4.9 mg/kg and

1.8 mg/L, respectively, and the blood ethanol concentration was 170 mg/100 mL *(66)*. Licata et al. described a homicide involving ketamine. The liver ketamine was 6.6 mg/kg and blood 27.4 mg/L *(67)*.

5.4. Cocaine

Baselt has summarized 19 cases where liver cocaine concentrations were measured as 0.1–20 mg/kg (average 4.2), with corresponding blood concentrations of 0.9–21 mg/L (average 5.3) *(13,36,68–75)*.

Bailey reported a series of 66 homicide cases and accidental overdoses. The liver concentrations ranged from 0 to 33 mg/kg for cocaine (average 0.96, SD = 4.59, N = 66) and 0 to 41 mg/kg for benzoylecgonine (average 5.9, SD = 11.59, N = 25). The corresponding blood values were 0-7.4 mg/L (average 0.46, SD = 1.40) and 0-14.4 mg/L (average 2.75, SD = 4.06) for cocaine and benzoylecgonine, respectively *(29)*.

Spiehler measured cocaine and benzoylecgonine concentrations in blood, liver, and brain in two series of cases, where the death was attributed to cocaine (37 cases) and where it was considered incidental (42 cases). In the cocaine fatality group, liver concentrations ranged from 0 to 393 mg/kg for cocaine (average 6.7) and 1.3 to 87 mg/kg for benzoylecgonine (average 21.3), the corresponding blood values were 0.04 to 31 mg/L (average 4.6) and 0.74 to 31 mg/L (average 7.8) for cocaine and benzoylecgonine, respectively. The second group with incidental cocaine findings had liver concentrations that ranged from 0 to 1.6 mg/kg for cocaine (average 0.08) and 0 to 10 mg/kg for benzoylecgonine (average 1.3), and the corresponding blood values were 0 to 0.5 mg/L (average 0.05) and 0 to 7.4 mg/L (average 0.88) for cocaine and BE, respectively *(76)*.

Apple and Roe reported six cases with liver concentrations of cocaine and benzoylecgonine ranging from 0 to 5.0 and 0.4 to 24.1 mg/kg, respectively; blood cocaine and benzoylecgonine concentrations ranged from 0 to 5.6 and 0.6 to 18.7 mg/L, respectively *(77)*.

Cingolani has shown that in formalin-preserved tissue, benzoylecgonine can be measured with recovery in fixed tissues averaging 12.31% from liver and 84.47% in formalin from the liver. No cocaine was detected in the liver *(78)*.

Shimomura examined liver, brain, blood, and urine specimens obtained from 15 postmortem cases for methyl ecgonidine, ecgonidine, cocaine and benzoylegonine. The median concentration, and range, for cocaine in blood was 0.012 (0–0.088) mg/L and for liver 0.057 (0–0.503) mg/kg, and for benzoylecgonine in blood 0.458 (0.030–2.071) mg/L and liver 0.821 (0.045–4.98) mg/kg *(79)*.

5.5. Cannabinoids

Because of the high fat solubility and low doses usually involved, liver is not particularly suitable for the measurement of delta-9-tetrahydrocanabinol and the carboxy–THC metabolite. However, there is a single fatal overdose recorded in the literature with a liver THC concentration of 38 mg/kg! The THC was estimated semi-quantitatively by thin-layer chromatography (TLC) and photoelectric densitometry *(80)*. No blood concentration was reported.

6. ADVANTAGES AND DISADVANTAGES

Liver is a specimen easily obtained at autopsy in sufficient quantities for comprehensive testing. It is generally considered the preferred specimen when blood is unavailable. Disadvantages of utilizing liver include the necessity to produce an homogenate prior to extraction, matrix effects observed with routine analytical techniques, and lack of a database to aid the interpretation of drug concentrations. The literature is deficient in describing the probable non-homogeneity in liver sampling. However, drug concentrations measured in liver are generally within the detection range of the analytical methods currently utilized in forensic toxicology.

This chapter has summarized the literature regarding the detection and measurement of typical drugs-of-abuse in liver. Corresponding blood levels have been included and so in this way may serve as the beginning of a database to evaluate the meaning of drug concentrations in this biological specimen.

REFERENCES

1. Pounder DJ, Adams E, Fuke C, Langford AM. Site to site variability of postmortem drug concentrations in liver and lung. *J Forensic Sci* 1996; 41(6):927–932.
2. Pounder DJ, Fuke C, Cox DE, Smith D, Kuroda N. Postmortem diffusion of drugs from gastric residue: an experimental study. *Am J Forensic Med Pathol* 1996; 17(1):1–7.
3. Parker JM, Winek CL, Shanor SP. Post-mortem changes in tissue levels of sodium secobarbital. *Clin Toxicol* 1971; 4(2):265–272.
4. Hilberg T, Bugge A, Beylich KM, Ingum J, Bjorneboe A, Morland J. An animal model of postmortem amitriptyline redistribution. *J Forensic Sci* 1993; 38(1): 81–90.
5. Hilberg T, Morland J, Bjorneboe A. Postmortem release of amitriptyline from the lungs; a mechanism of postmortem drug redistribution. *Forensic Sci Int* 1994; 64(1):47–55.
6. Osselton MD. The release of basic drugs by the enzymic digestion of tissues in cases of poisoning. *J Forensic Sci Soc* 1978; 17(2–3):189–194.

7. Osselton MD, Shaw IC, Stevens HM. Enzymic digestion of liver tissue to release barbiturates, salicylic acid and other acidic compounds in cases of human poisoning. *Analyst* 1978; 103(1232):1160–1164.

8. Matuszewski BK, Constanzer ML, Chavez-Eng CM. Matrix effect in quantitative LC/MS/MS analyses of biological fluids: a method for determination of finasteride in human plasma at picogram per milliliter concentrations. *Anal Chem* 1998; 70(5):882–889.

9. Muller C, Schafer P, Stortzel M, Vogt S, Weinmann W. Ion suppression effects in liquid chromatography-electrospray-ionisation transport-region collision induced dissociation mass spectrometry with different serum extraction methods for systematic toxicological analysis with mass spectra libraries. *J Chromatogr B Analyt Technol Biomed Life Sci* 2002; 773(1):47–52.

10. Shen JX, Motyka RJ, Roach JP, Hayes RN. Minimization of ion suppression in LC-MS/MS analysis through the application of strong cation exchange solid-phase extraction (SCX-SPE). *J Pharm Biomed Anal* 2005; 37(2):359–367.

11. Kudo K, Ishida T, Nishida N et al. Simple and sensitive determination of free and total morphine in human liver and kidney using gas chromatography-mass spectrometry. *J Chromatogr B Analyt Technol Biomed Life Sci* 2006; 830(2): 359–363.

12. Cingolani M, Froldi R, Mencarelli R, Mirtella D, Rodriguez D. Detection and quantitation of morphine in fixed tissues and formalin solutions. *J Anal Toxicol* 2001; 25(1):31–34.

13. Baselt RC. *Disposition of Toxic Drugs and Chemical in Man.* 7th ed. Foster City, CA: Biomedical Publications, 2004.

14. Finkle B. Vio-Dex (amphetamine) death. *Bull Int Assoc Forensic Toxicol* 1967; 4(4):4.

15. Finkle B. Amphetamine death. *Bull Int Assoc Forensic Toxicol* 1970; 7(3):3–4.

16. Orrenius S, Maehly AC. Lethal amphetamine intoxication. A report of three cases. *Z Rechtsmed* 1970; 67(3):184–189.

17. Richards HG, Stephens A. Sudden death associated with the taking of amphetamines by an asthmatic. *Med Sci Law* 1973; 13(1):35–38.

18. Van HF, Heyndrickx A, Timperman J. Report of a human fatality due to amphetamine. *Arch Toxicol* 1974; 32(4):307–312.

19. Adjutantis G, Coutselinis A, Dimopoulous G. Fatal intoxication with amphetamines (a case report). *Med Sci Law* 1975; 15(1):62–63.

20. Bailey DN. Amphetamine sulfate. *Bull Int Assoc Forensic Toxicol* 1976; 12(3):6.

21. Dams R, De Letter EA, Mortier KA et al. Fatality due to combined use of the designer drugs MDMA and PMA: a distribution study. *J Anal Toxicol* 2003; 27(5):318–322.

22. Cravey RH, Baselt RC. Methamphetamine poisoning. *J Forensic Sci Soc* 1968; 8(2):118–120.

23. Randall C. Baselt. *Disposition of Toxic Drugs and Chemicals in Man.* 7th ed. Foster City, CA: Biomedical Publications, 2004.

24. Kojima T, Une I, Yashiki M. CI-mass fragmentographic analysis of methamphetamine and amphetamine in human autopsy tissues after acute methamphetamine poisoning. *Forensic Sci Int* 1983; 21(3):253–258.

25. Kojima T, Une I, Yashiki M, Noda J, Sakai K, Yamamoto K. A fatal methamphetamine poisoning associated with hyperpyrexia. *Forensic Sci Int* 1984; 24(1): 87–93.

26. Moore KA, Daniel JS, Fierro M, Mozayani A, Poklis A. The detection of a metabolite of alpha–benzyl-N-methylphenethylamine synthesis in a mixed drug fatality involving methamphetamine. *J Forensic Sci* 1996; 41(3):524–526.

27. Sato Y, Kondo T, Takayasu T, Ohshima T. [Detection of methamphetamine in a severely burned cadaver–a case report]. *Nippon Hoigaku Zasshi* 2000; 54(3): 420–424.

28. Logan BK, Weiss EL, Harruff RC. Case report: distribution of methamphetamine in a massive fatal ingestion. *J Forensic Sci* 1996; 41(2):322–323.

29. Bailey DN, Shaw RF. Cocaine- and methamphetamine-related deaths in San Diego County (1987): homicides and accidental overdoses. *J Forensic Sci* 1989; 34(2):407–422.

30. Cimbura G. 3,4-methylenedioxyamphetamine (MDA): analytical and forensic aspects of fatal poisoning. *J Forensic Sci* 1972; 17(2):329–333.

31. Kier L. MDA poisoning. *Bull Int Assoc Forensic Toxicol* 1972; 9(2):5.

32. Reed D, Cravey RH, Sedgwick PR. A fatal case involving methylenedioxyamphetamine. *Clin Toxicol* 1972; 5(1):3–6.

33. Fiorese F. Fatal MDA poisoning. *Bull Int Assoc Forensic Toxicol* 1974; 10(2): 15–16.

34. Lukaszewski T. 3,4-methylenedioxyamphetamine overdose. *Clin Toxicol* 1979; 15(4):405–409.

35. Poklis A, Mackell MA, Drake WK. Fatal intoxication from 3,4-methylenedioxyamphetamine. *J Forensic Sci* 1979; 24(1):70–75.

36. Gottschalt LA, Cravey RH. *Toxicological and Pathological Studies on Psychoactive Drugs-Involved Deaths.* Davis, CA: Biomedical Publications, 1980.

37. De Letter EA, Clauwaert KM, Lambert WE, Van Bocxlaer JF, De Leenheer AP, Piette MH. Distribution study of 3,4-methylenedioxymethamphetamine and 3,4-methylenedioxyamphetamine in a fatal overdose. *J Anal Toxicol* 2002; 26(2): 113–118.

38. Sticht G, Pluisch F, Bierhoff E, Kaferstein H. [Fatal outcome of Ecstasy overdose]. *Arch Kriminol* 2003; 211(3-4):73–80.

39. Cimbura G. PMA deaths in Ontario. *Can Med Assoc J* 1974; 110(11):1263–1267.

40. Felgate HE, Felgate PD, James RA, Sims DN, Vozzo DC. Recent paramethoxyamphetamine deaths. *J Anal Toxicol* 1998; 22(2):169–172.

41. Tucker R. Fatal PMA (p-methoxyamphetamine) poisoning. *Bull Int Assoc Forensic Toxicol* 1973; 9(3):15.

42. Felby S, Christensen H, Lund A. Morphine concentrations in blood and organs in cases of fatal poisoning. *Forensic Sci* 1974; 3(1):77–81.

43. Cravey RH, Reed D. The distribution of morphine in man following chronic intravenous administration. *J Anal Toxicol* 1977; 1:166–167.

44. Cravey RH. An unusually high blood morphine concentration in a fatal case. *J Anal Toxicol* 1985; 9:237.

45. Chan SC, Chan EM, Kaliciak HA. Distribution of morphine in body fluids and tissues in fatal overdose. *J Forensic Sci* 1986; 31(4):1487–1491.

46. F, Hashimoto Y. Distribution of free and conjugated morphine in body fluids and tissues in a fatal heroin overdose: is conjugated morphine stable in postmortem specimens? *J Forensic Sci* 1997; 42(4):736–740.

47. Spiehler V, Brown R. Unconjugated morphine in blood by radioimmunoassay and gas chromatography/mass spectrometry. *J Forensic Sci* 1987; 32(4):906–916.

48. Spiehler VR. Computer-assisted interpretation in forensic toxicology: morphine-involved deaths. *J Forensic Sci* 1989; 34(5):1104–1115.

49. Kerrigan S, Honey D, Baker G. Postmortem morphine concentrations following use of a continuous infusion pump. *J Anal Toxicol* 2004; 28(6):529–532.

50. Lewis RJ, Johnson RD, Hattrup RA. Simultaneous analysis of thebaine, 6-MAM and six abused opiates in postmortem fluids and tissues using Zymark automated solid-phase extraction and gas chromatography-mass spectrometry. *J Chromatogr B Analyt Technol Biomed Life Sci* 2005; 822(1-2):137–145.

51. Nakamura GR, Way EL. Determination of morphine and codeine in post-mortem specimens. *Anal Chem* 1975; 47(4):775–778.

52. Nakamura GR, Griesemer EC, Noguchi TT. Antemortem conversion of codeine to morphine in man. *J Forensic Sci* 1976; 21(3):518–524.

53. Sedgwick P. Driving under the influence? *Bull Int Assoc Forensic Toxicol 1973*; 9(3):12.

54. Cravey RH. *Coutroom Toxicology*. New York: Matthew Bender, 1996.

55. Armstrong EJ, Jenkins AJ, Sebrosky GF, Balraj EK. An unusual fatality in a child due to oxycodone. *Am J Forensic Med Pathol* 2004; 25(4):338–341.

56. Levine B, Moore KA, ronica-Pollak P, Fowler DF. Oxycodone intoxication in an infant: accidental or intentional exposure? *J Forensic Sci* 2004; 49(6):1358–1360.

57. Walls HC, Slightom L. Hydromorphone death. *Bull Int Assoc Forensic Toxicol* 1976; 12(3):7–8.

58. Levine B, Saady J, Fierro M, Valentour J. A hydromorphone and ethanol fatality. *J Forensic Sci* 1984; 29(2):655–659.

59. Manning T, Bidanset JH, Cohen S, Lukash L. Evaluation of the Abuscreen for methadone. *J Forensic Sci* 1976; 21(1):112–120.

60. Siek TJ. The analysis of meperidine and normeperidine in biological specimens. *J Forensic Sci* 1978; 23(1):6–13.

61. Reynolds PC. Clinical and forensic experiences with phencyclidine. *Clin Toxicol* 1976; 9(4):547–552.

62. Noguchi TT, Nakamura GR. Phencyclidine-related deaths in Los Angeles County, 1976. *J Forensic Sci* 1978; 23(3):503–507.

63. Caplan YH, Orloff KG, Thompson BC. Detection of phencyclidine in medical Examiner's cases. *J Anal Toxicol* 1979; 3:47–52.

64. Cravey RH, Reed D, Ragle JL. Phencyclidine -related deaths: a report of nine fatal cases. *J Anal Toxicol* 1979; 3:199–201.

65. Peyton SH, Couch AT, Bost RO. Tissue distribution of ketamine: two case reports. *J Anal Toxicol* 1988; 12(5):268–269.

66. Moore KA, Kilbane EM, Jones R, Kunsman GW, Levine B, Smith M. Tissue distribution of ketamine in a mixed drug fatality. *J Forensic Sci* 1997; 42(6): 1183–1185.

67. Licata M, Pierini G, Popoli G. A fatal ketamine poisoning. *J Forensic Sci* 1994; 39(5):1314–1320.
68. McCurdy HH, Jones JK. Unusual cocaine death. *Bull Int Assoc Forensic Toxicol 1973*; 9(3):5.
69. Price KR. Fatal cocaine poisoning. *J Forensic Sci Soc* 1974; 14(4):329–333.
70. Griffin BR. Cocaine death. *Bull Int Assoc Forensic Toxicol* 1975; 11(1):6.
71. Lundberg GD, Garriott JC, Reynolds PC, Cravey RH, Shaw RF. Cocaine-related death. *J Forensic Sci* 1977; 22(2):402–408.
72. Prouty R. A unique cocaine fatality. American Academy of Forensic Sciences Annual Meeting, San Diego, CA, 1977.
73. DiMaio V, Garriott JC. Four deaths due to intravenous injection of cocaine. *Forensic Sci Int* 1978; 12(2):119–125.
74. Bednarczyk LR, Gressmann EA, Wymer RL. Two cocaine-induced fatalities. *J Anal Toxicol* 1980; 4(5):263–265.
75. Poklis A, Mackell MA, Graham M. Disposition of cocaine in fatal poisoning in man. *J Anal Toxicol* 1985; 9(5):227–229.
76. Spiehler VR, Reed D. Brain concentrations of cocaine and benzoylecgonine in fatal cases. *J Forensic Sci* 1985; 30(4):1003–1011.
77. Apple FS, Roe SJ. Cocaine-associated fetal death in utero. *J Anal Toxicol* 1990; 14(4):259–260.
78. Cingolani M, Cippitelli M, Froldi R, Gambaro V, Tassoni G. Detection and quantitation analysis of cocaine and metabolites in fixed liver tissue and formalin solutions. *J Anal Toxicol* 2004; 28(1):16–19.
79. Shimomura ET, Hodge GD, Paul BD. Examination of postmortem fluids and tissues for the presence of methylecgonidine, ecgonidine, cocaine, and benzoylecgonine using solid-phase extraction and gas chromatography-mass spectrometry. *Clin Chem* 2001; 47(6):1040–1047.
80. Tewari SN, Sharma JD. Detection of delta-9-tetrahydrocannabinol in the organs of a suspected case of cannabis poisoning. *Toxicol Lett* 1980; 5(3-4):279–281.

Chapter 10

Drugs-of-Abuse Testing in Brain

Thomas Stimpfl

Summary

The majority of articles concerning drugs-of-abuse in human brain specimens were published in the 1980s. They focused primarily on opiates, cocaine, and cocaine metabolites – specifically, the interpretation of data when determining time intervals between administration of the drug and death and in determining the role of these drugs in the cause of death itself. This chapter presents an overview of those publications, as well as developments in techniques for sample preparation, automation, and detection, which, when combined with the routine use of stable isotope internal standards, promise more comparable results and could lay the foundation for a data collection of reliable reference values. Moreover, recently developed immunohistochemical techniques could pave the way for the systematic investigation of the function of drugs-of-abuse in specific substructures of the brain.

Key Words: Postmortem forensic toxicology, drugs-of-abuse, brain.

1. STRUCTURE AND COMPOSITION OF THE HUMAN BRAIN

A brief overview of the anatomy of the brain is provided, focusing primarily on the regional distribution of the binding sites of drugs-of-abuse in the human brain whenever published data is available. Generally, the brain can be divided into the following parts: telencephalon, diencephalon, mesencephalon (or midbrain), pons and cerebellum, and medulla oblongata. The telencephalon is composed of the two large cerebral hemispheres containing the basal ganglia. The diencephalon is situated between the cerebral hemispheres

From: *Forensic Science and Medicine: Drug Testing in Alternate Biological Specimens*
Edited by: A. J. Jenkins © Humana Press, Totowa, NJ

and is divided into the thalamus and the hypothalamus. The remaining parts of the brain are grouped together to form the brainstem: The mesencephalon and the medulla are divided by the pons, which is separated from the cerebellum by a cavity called the fourth ventricle. The border between the medulla oblongata and the spinal cord is defined by the pyramidal decussation (in the height of the foramen magnum). These anatomical subdivisions are interconnected, creating the necessity to define various substructures that describe spatial relationships (nuclei, tracts, etc.). A detailed illustration would be beyond the scope of this chapter and can be found elsewhere *(1)*.

The central nervous system (CNS) is comprised of gray matter (mainly composed of nerve cell bodies), white matter (mainly composed of white axon fibers), and glial cells. About 67–84% of the total weight of the brain is water. The remaining 16–33% is mainly proteins and lipoids – in nearly equal amounts. A structural protein unique to the CNS is neurokeratin, and the most common amino acid found in the brain is glutamic acid. The concentration of lipoids is higher in white matter than in gray matter. The lipoid fraction contains phosphatides (cephalines, lecithines, and sphingomyelines), cerebrosides (glycolipids), cholesterol, and gangliosides (Table 1). The most common fatty acids found in the brain are unsaturated C20- and C22-fatty acids, palmitic acid, stearic acid, oleic acid, arachidonic acid, docosanoic acid, and C24-fatty acids *(2,3)*. Phosphatides and cerebrosides tend to form viscous, stringy solutions in water, which can create problems during the extraction of brain specimens.

Because most drugs-of-abuse establish their effects through specific receptors in the brain, their regional distribution is of high interest. The location of binding sites for drugs-of-abuse can be studied by quantitative autoradiography in vitro. After proven successful in animal studies, this method was further applied to human brain tissue. Biegeon et al. performed a binding study with nanomolar and micromolar (representative for behaviorally active doses) concentrations of cocaine in brain sections of three drug-free subjects.

For low concentrations of tritiated cocaine, the highest density of binding sites was found in the basal ganglia (caudate and putamen, also referred to as

Table 1
Lipoids Found in the Human Brain (% of Wet Weight) (2,3)

	White matter	Gray matter
Phosphatides	6–7	3–3.5
Cerebrosides	4–4.5	0.5–1
Cholesterol	4	1

the striatum). The thalamus showed moderate density, and density was low in the cortex and hippocampus (parts of the telencephalon). High concentrations of tritiated cocaine showed more homogeneous binding throughout the brain *(4)*.

A study to determine the regional distribution of the three different opiate receptor types (mu, delta, and kappa) in human brain tissue was performed by Maurer et al. *(5)* Opiate-binding sites were generally predominant in grey matter and almost absent in white matter. A very high density of μ-opiate receptors – through which morphine exerts most of its pharmacological actions – was found in the brainstem (area tegmentalis ventralis, grisum centrale mesencephali, inferior colliculus and nucleus interpeduncularis); moderate density was found in the molecular layer of the cerebellar cortex (also a part of the brainstem) and the neocortex (laminae I–V) and hippocampus (gyrus dentatus) of the telencephalon *(5)*. More recently, quantitative autoradiography detected μ-opiate receptors in the telencephalon, with high density in the laminae I–III of the neocortex, and the nucleus caudatus; moderate density in the laminae III–IV of the neocortex, the nucleus basalis of Meynert, and the corpus amygdaloideum; and low density in the laminae V–VI of the neocortex and the claustrum. Moderate density was also found in the thalamus (diencephalon) and the cerebellum (brainstem). Low density was found in the hypothalamus (diencephalon) *(6)*. Gabilondo et al. demonstrated that in the postmortem brain of heroin addicts, no apparent alterations in the densities and affinities of μ-opiate receptors in various brain regions could be observed when compared to the control group. Brain α_2-adrenoceptor densities, however, appeared to be down-regulated during opiate dependence in humans *(7)*.

The highest density of specific cannabinoid receptors was found in the telencephalon – in the basal ganglia (substantia nigra pars rediculata and globus pallidus) and in the hippocampus – and in the cerebellum (brainstem). The very low density in the brainstem areas controlling cardiovascular and respiratory functions may explain why there are no reports of fatal cannabis intoxication in humans *(8)*.

2. KINETICS OF DRUG TRANSFER

Upon absorption into the body, the inherent properties of drugs-of-abuse (such as molecular size, lipid solubility, degree of ionization, and plasma protein binding) determine their capacity to cross capillary walls and penetrate into cells. Only the unbound fraction circulating in the blood has the ability to penetrate the blood–brain barrier, and the permeability of this barrier determines the amount of drug reaching the site of action in the brain.

Oldendorf et al. studied the uptake of different opiates in the brains of rats, and predicted rates of penetration through the blood–brain barrier *(9)*. In the extremely short time frame of 15 s between bolus injection into the carotid artery and decapitation of the rats, the permeability of the blood–brain barrier for morphine was less than measurable. The uptake of codeine, heroin, and methadone was 24, 68, and 42%, respectively. Bjoerkman et al. characterized the cerebral uptake of morphine in pigs by measuring changes in the arterio-venous plasma concentration gradient over the brain. Over a much longer observation period (90 min) than in the previously mentioned study, it was shown that morphine uptake in the brain reached a maximum approximately 3 min after the start of the infusion before changing to a slow, steady release *(10)*. These results are in agreement with a study performed by Mullis et al. *(11)*, who determined the half-life of morphine in rat brain tissue to be biphasic, with an initial half-life of approximately 2 h followed by a slower half-life of about 5 h between 4 and 48 h after a single subcutaneous injection. It could be expected that the redistribution of morphine back into blood would also be retarded by low permeability of the blood–brain barrier. Indeed, unconjugated morphine persisted in the rat brains in nanomolar concentrations for at least 24 h after a single analgesic dose *(11)*. Wu et al. demonstrated that the brain uptake of morphine-6-glucuronide was 32 times lower than that of morphine *(12)*. This reduced blood–brain barrier permeability is consistent with the much lower lipid solubility of morphine-6-glucuronide relative to morphine.

Som et al. demonstrated the rapid, intense uptake of cocaine in rat brains after a pharmacological dose was injected intravenously *(13)*. Mule et al. confirmed the rapid entry of cocaine into the brain matrix and demonstrated that this drug also leaves the brain rapidly. The penetration of benzoylecgonine was considerably slower; that would be expected due to its high polarity, lower lipid solubility, and the partition coefficient of 0.15 *(14)*. Furthermore, the blood–brain barrier seems to have a limiting effect on the access and accumulation of tetrahydrocannabinol (THC) in the brain *(15)*.

After penetration of the blood–brain barrier, biotransformation can take place within the brain. Examples include heroin, which largely survives the 10–15 s required to reach and enter the brain after intravenous injection and is then rapidly hydrolyzed into 6-acetylmorphine and morphine; another example would be the hydrolysis of cocaine into benzoylecgonine.

Although drugs-of-abuse are distributed throughout the brain, for several there appears to be an uneven distribution. Therefore, the brain cannot be regarded as a single pharmacokinetic compartment. Morphine concentrations in the hypothalamus and in the hippocampus of rats, for example, were signifi-cantly higher than those found in the midbrain-thalamus and the striatum *(16)*. The highest concentration of cocaine in rat brains was found in the cortex, followed by the striatum and the cerebellum *(13)*.

Animal studies, however, are not always representative of human conditions, and in contrast to the previously mentioned animal study, distribution of cocaine in postmortem human brain appears to be uniform. These findings and the question as to whether there is a correlation between the concentration of drugs-of-abuse in different brain regions and the proposed sites of action will be discussed in the section on Interpretational Issues.

3. SAMPLE PREPARATION PROCEDURES AND INSTRUMENT TESTING METHODOLOGIES

Appropriate sample preparation is one of the most important pre-requisites for the successful identification and quantification of drugs-of-abuse in brain specimens. It is a multi-step process consisting of sampling, sample pre-treatment, sample extraction, chromatographic separation, and detection. A high level of accuracy at each step is essential to ensure the correct interpretation of results.

3.1. Sampling

Sampling of the brain is crucial and demands particular care by the professional conducting the autopsy. Specimen amount, as well as the area of the brain from which the specimen is taken, must be carefully documented. To minimize specimen degradation, specimens should be analyzed as soon as possible and kept frozen; unstable drugs-of-abuse (e.g., cocaine) must be properly preserved.

3.2. Sample Pre-Treatment

The goal of pre-treatment is to make drugs-of-abuse accessible for the extraction process that is to follow while simultaneously removing interfering compounds. Tissue specimens such as brain specimens require careful homogenization (e.g., using an electric-powered blender or equivalent device) to obtain a sample that is suitable for extraction. This process can be followed by protein precipitation, but potential irreversible loss of significant compounds due to adsorption and occlusion must be considered *(17)*. If protein precipitation cannot be avoided, possible adsorption losses must be compensated for by adding stable isotope internal standards in the expected concentrations of the targeted drugs-of-abuse. Moreover, digestion of the homogenates with lipases or proteases gives rise to increased background due to the co-extraction of an increased number of artifacts *(18)*. Centrifugation prior to extraction should be performed in a cooled centrifuge so as to prevent any loss of analytes from the warming of the sample.

Generally, a major problem in the work-up procedure of tissue specimens is the impossibility of determining the actual extraction efficiency and hence the absolute content of drugs in the tissue specimen. Nevertheless, Drummer et al. pointed out that when suitable precautions are taken, there is little evidence to suggest that extraction efficiencies of drugs from solid tissues are likely to be much worse than with fluid specimens *(19)*.

Therefore, internal standards – with chemical and physical properties as similar as possible to the analytes – should be added to the sample at the earliest conceivable stage, in any event before buffering and extraction of the sample, to compensate for possible loss of target compounds throughout the entire sample preparation procedure.

Today, stable isotope internal standards are generally recommended and should be added in similar concentrations to the expected drugs-of-abuse, but they require assays based on mass spectrometry (MS). Whenever available, certified reference materials should be used as controls. However, in many forensic cases, the matrix is unique (e.g., a decomposed brain specimen); in these cases, the control can never reach the ideal of being identical to the unknown specimen, and quantitative results must be interpreted cautiously.

3.3. Sample Extraction

Sophisticated extraction procedures are needed for the isolation and concentration of targeted drugs-of-abuse from the complex matrix of brain specimens. At the same time, as many interferences as possible originating from the specimen should be excluded. In the past, liquid–liquid extraction (LLE) and, more recently, solid-phase extraction (SPE) have been applied to reach those goals. Today, SPE procedures are often favored because automated devices for sample extraction are available, which minimize systematic errors and human-caused variances and improve reproducibility *(20)*.

3.3.1. LIQUID–LIQUID EXTRACTION

The advantages of LLE are its wide dynamic range and that it is based on strictly defined thermodynamic relationships, which simplifies method development. LLE can be performed at pH values optimized for the targeted drugs-of-abuse, and basic or acidic back-extraction can be applied to remove interferences. Because LLE is frequently combined with protein precipitation, extraction yields are often highly variable due to adsorption and occlusion. Moreover, during extraction, stable emulsions occur and multiple interferences are co-extracted from the brain matrix. To prevent emulsions and eliminate the need for centrifugation, the aqueous phase has been immobilized onto inert support materials, which improves both extraction efficiency and reproducibility *(21)*. However, the applicability of this technique for brain specimens is limited, because homogenization of the potential infectious specimens with the support material is difficult. The need for an inert carrier to support the sample during extraction also seems to be the major drawback for the application of both accelerated solvent extraction (ASE) and supercritical fluid extraction (SFE) to brain specimens.

In the past, several procedures for LLE of animal and human brain specimens have been published for opiates *(21–32)*, for methadone *(33)*, for cocaine and metabolites *(34–45)*, and for amphetamines *(46–51)*.

3.3.2. SOLID-PHASE EXTRACTION

In SPE, analytes are isolated from the aqueous matrix onto a solid sorbent followed by selective washing – to remove interferences from the biological matrix – and elution with appropriate solvents. The high extraction efficiency allows for small specimen amounts, which is crucial in forensic cases because often only a limited sample is available. Moreover, SPE offers the advantage of reduced solvent consumption, decreased evaporation volumes, as well as reduced environmental waste and personnel exposure to harmful substances. Furthermore, there is no problem with emulsions, and each step of the extraction process can be automated. This increases both reproducibility and laboratory turnover. The reliability of SPE has generally improved since manufacturers began to implement quality control systems for the sorbents. Although method development of SPE is more complex than that for LLE, SPE procedures covering a broad range of drugs-of-abuse have been developed primarily for body fluids such as urine or blood. Scheurer et al. reviewed the few existing SPE procedures for tissue specimens and concluded that the problems unique to brain tissue were not insurmountable and that existing procedures published for animal brain would be equally applicable to human tissue *(52)*.

Browne et al. reported that lipids present in brain homogenates cause C18- and C8-extraction columns to become blocked as a result of their interaction with the lipophilic packing materials and showed that polar C2-columns worked better *(18)*. Another approach to this problem would be the digestion of the brain specimens with triacylglycerol lipase prior to extraction *(53)*. The main drawback of SPE for postmortem brain specimens is the inability of tissue homogenates to pass easily through tightly packed bonded cartridges. To cope with the increased viscosity of brain specimens, high-flow SPE columns were developed, but the high speed at which the samples pass through the packing material reduces the contact time between drug and sorbent, resulting in a lower extraction efficiency compared to procedures using more common tightly packed sorbents *(54)*. Silica-based bonded sorbents and polystyrene–divinylbenzene copolymers have been used in the extraction of human brain specimens. Procedures have been published for opiates *(55–58)*, for cocaine and metabolites *(18,53,54,59–62)*, for amphetamines and analogs *(63–65)*, and for cannabinoids *(45)*. For the extraction of brain specimens, polystyrene resins offer a major advantage compared to other sorbents. The porous structure enables the exclusion of micelles containing monoglycerides, diglycerides, and phospholipids. These micelles are produced by dispersion during homogenization of the specimen and, because they are prevented from adsorption, lipoid interferences are reduced in the resulting extract *(58)*.

3.4. Chromatographic Separation and Detection (Identification)

To enable accurate interpretation of brain drug concentrations in determining cause of death, these substances must be identified and the quantitative results must be sufficient with regard to sensitivity and precision. This is the reason why methods such as immunoassay, fluorometric assay, and thin-layer chromatography, although applied in the past, are no longer recommended. Differential radioimmunoassay, for instance, allows for the separate determination of a targeted drug-of-abuse and its metabolites, yet other cross-reacting substances cannot be ruled out. Fluorometric assays are limited by quenching interferences. Thin-layer chromatography lacks sensitivity and does not produce precise quantitative results.

Methods such as high-performance liquid chromatography (HPLC) with electrochemical, ultraviolet, and fluorescence detection, as well as gas chromatography (GC) with electron capture, flame ionization, and nitrogen phosphorus detection, have been successfully applied for the detection of drugs-of-abuse in brain specimens. To obtain reliable quantitative results with these procedures, internal standards with chemical and physical properties as similar as possible to the analytes are needed. Because internal standards that meet all these demands are not easily available, quantitative results must be interpreted with caution. For the majority of drugs-of-abuse, stable isotope analogs are available and can be used as "ideal" internal standards, provided MS is utilized. If added at the earliest possible stage of the work-up procedure in a similar concentration to the targeted drugs-of-abuse, stable isotope internal standards correct any loss of the target compound during the whole process of sample preparation and quantification. In the heterogeneous brain matrix, MS, in combination with chromatographic techniques, offers the highest selectivity, sensitivity, and most precise quantitative results. Therefore, combinations of GC and MS or LC–MS have developed into the standard techniques for the detection of drugs-of-abuse today and should be applied whenever possible. For GC–MS, appropriate derivatization may be necessary to achieve sufficient volatility for drugs-of-abuse. At the same time, care must be taken to prevent injection port temperatures that are too high, leading to the production of artifacts (e.g., anhydroecgonine methyl ester). These problems do not exist when LC–MS is applied; instead, matrix effects – which interfere with the ionization process – must be taken into consideration *(66)*.

4. DRUGS DETECTED IN BRAIN

Morphine, 6-acetylmorphine, cocaine, and cocaine metabolites have been detected in human brain specimens, and issues concerning the interpretation of these results have been discussed in the literature. In contrast, little data on the concentration of methadone, amphetamine, methamphetamine, 3,4-methylenedioxymethamphetamine (MDMA), 3,4-methylenedioxyamphetamine (MDA), 3,4-methylenedioxyethylamphetamine (MDEA), cannabinoids, and phencyclidine (PCP) in human brain specimens are available, and interpretational issues are unaddressed. For all other drugs-of-abuse, there is no data available regarding human brain specimens. The majority of the publications were written in the 1980s, and numerous extraction and detection procedures were applied. Moreover, different amounts of specimen were used and only a few publications gave the precise region of the brain assayed and the time between death and sampling. Data, therefore, are not directly comparable, and interpretation or conclusions should be drawn very cautiously. To provide an overview, the available literature is summarized in Tables 2–7.

5. INTERPRETATIONAL ISSUES

5.1. Opiates

In the human body, heroin is rapidly deacetylated to 6-acetylmorphine and then to morphine. In vitro studies have demonstrated that human liver homogenates possessed the greatest hydrolytic activity and human brain tissue the least *(76)*. Heroin has not been detected in brain specimens though 6-acetylmorphine levels in the brain were found to be substantially higher than in blood, liver, lung, and kidney specimens *(56)*. Sticht et al. detected 6-acetylmorphine in all brain specimens of heroin intoxication cases where the proportion of unconjugated morphine to total morphine in the blood exceeded 45%. The presence of 6-acetylmorphine in the brain denotes heroin consumption and suggests that consumption took place within 4.5 h of the time of death *(31)*.

Morphine is further metabolized by conjugation, which results in the formation of morphine-glucuronides (morphine-6-glucuronide is physiologically active), and by N-demethylation to normorphine. Based on lower lipid

Table 2
Opiates Detected in Human Brain

Reference	Number of cases	Brain regions defined	Extraction technique	Detection technique	Deuterated standards
Robinson et al. *(22)*	2	–	LLE	GC–FID	n/a
Richards et al. *(67)*	52	–	n.d.	Fluorometric	n/a
Reed et al. *(23)*	2	–	LLE	RIA and fluorometric	n/a
Spiehler et al. *(24)*	33	+	LLE	Fluorometric	n/a
Reed *(25)*	41	–	LLE	GC–MS	+
Ziminski et al. *(68)*	13	–	n.d.	RIA	n/a
Pare et al. *(27)*	21	+	LLE	GC–FID	n/a
Spiehler et al. *(28)*	144	–	LLE	GC–MS	+
Vycudilik *(55)*	62	+	SPE	GC–MS	–
Pollak et al. *(69)*	1	+	SPE	GC–MS	–
Kintz et al. *(29)*	4	–	LLE	GC–NPD	n/a
Sticht et al. *(31)*	13	–	LLE	GC–MS	+
Goldberger et al. *(56)*	2	–	SPE	GC–MS	+
Moriya *(32)*	1	+	LLE	GC-MS	–
Heinemann et al. *(70)*	1	–	n.d.	n.d.	n.d.
Klingmann et al. *(57)*	1	–	SPE	LC–MS	+

–, no; +, yes; n/a, not applicable; n.d., not defined.

solubility, morphine-glucuronides show reduced blood–brain barrier permeability relative to morphine, which is supported by the findings of Reed et al., where morphine concentrations increased approximately 260% in blood specimens following hydrolysis, whereas brain hydrolysis did not increase the yield more than 25% *(25)*.

Table 3
Methadone Detected in Human Brain

Reference	Number of cases	Brain regions defined	Extraction technique	Detection technique	Deuterated standards
Norheim *(33)*	1	–	LLE	GC–MS	–
Ziminski et al. *(68)*	2	–	n.d.	RIA	n/a

–, no; +, yes; n/a, not applicable; n.d., not defined.

Table 4
Cocaine and Metabolites Detected in Human Brain

Reference	Number of cases	Brain regions defined	Extraction technique	Detection technique	Deuterated standards
Lundberg et al. *(34)*	3	–	LLE	GC	n/a
Chinn et al. *(35)*	1	–	LLE	GC–MS	+
Poklis et al. *(37)*	1	–	LLE	GC–NPD	n/a
Spiehler et al. *(38)*	83	+/–	LLE	GC–MS	+
Poklis et al. *(39)*	5	–	LLE	GC–NPD	n/a
Mittleman et al. *(40)*	2	–	LLE	GC–NPD	n/a
Morild et al. *(71)*	59	–	n.d.	n.d.	n.d.
Browne et al. *(18)*	3	+	SPE	HPLC-UV	n/a
Hime et al. *(41)*	3	–	LLE	GC–NPD	n/a
Hernandez et al. *(53)*	10	–	SPE	GC–MS	+
Heinemann et al. *(70)*	2	+/–	n.d.	n.d.	n.d.
Stichenwirth et al. *(59)*	1	+	SPE	GC–MS	–
Kalasinsky et al. *(60)*	15	+	SPE	GC–MS	+
Shimomura et al. *(61)*	15	–	SPE	GC–MS	+
Furnari et al. *(62)*	1	–	SPE	GC–MS	n.d.
Giroud et al. *(45)*	1	+	LLE	LC–MS	+

–, no; +, yes; n/a, not applicable; n.d., not defined.

Published data includes concentrations of total morphine (morphine and morphine glucuronides) as well as concentrations of unconjugated morphine in brain specimens, involving a wide concentration range, thus making interpretation difficult. The question of distribution of morphine within the brain has been addressed in different studies, but it is difficult to compare these results because different analytical procedures were utilized.

A study by Spiehler et al. analyzed of the following regions of the CNS: cerebellum, cerebral cortex, mid-brain, pons, medulla, spinal cord, and cerebrospinal fluid (CSF). Results were presented for 33 cases, where death was mainly attributed to intravenous administration of heroin. Although total morphine concentrations in the brain were found to vary from region to region in each case, no consistent overall pattern was found, and there was also no correlation of total morphine distribution between brain regions. Moreover, no correlation between total morphine concentrations in brain fluid (CSF) and brain tissue could be established. In cases of slow death (death occurring 2–48 h after administration of heroin), total morphine concentrations in the CSF, spinal cord, and medulla tended to be elevated. The authors concluded that

Table 5
Amphetamines Detected in Human Brain

Reference	Number of cases	Brain regions defined	Extraction technique	Detection technique	Deuterated standards
Kojima et al. *(46)*	1	–	LLE	GC–MS	–
Kojima et al. *(47)*	1	–	LLE	GC–MS	–
Hara et al. *(49)*	6	–	LLE	GC–MS	+
Rohrig et al. *(72)*	2	–	n.d.	GC–FID	n/a
Katsumata et al. *(73)*	1	+	n.d.	GC–MS	n.d.
Meyer et al. *(63)*	1	–	SPE	GC–FID	n/a
Heinemann et al. *(70)*	1	–	n.d.	n.d.	n.d.
Weinmann et al. *(64)*	1	+	SPE	GC–MS	+
Fineschi et al. *(74)*	3	–	n.d.	GC–MS	–
Kalasinsky et al. *(50)*	14	+	LLE	GC–MS	n.d.
De Letter et al. *(51)*	1	+	LLE	HPLC – fluorescence detection	n/a
Garcia-Repetto et al. *(65)*	2	–	SPE	GC–NPD	n/a

–, no; +, yes; n/a, not applicable; n.d., not defined.

there was a lack of correlation between distribution of morphine in the brain and proposed sites of action *(24)*.

In another study by Spiehler et al., 52 brain specimens (cerebral cortex) from cases of acute intoxication (defined in this study as death occurring less than 3 h after heroin or morphine injection) showed a mean total morphine concentration of 0.47 μg/g in the brain with a wide range of 0.03–3.4 μg/g. In 92 brain specimens where death occurred more than 3 h after administration or the time interval was unknown, total morphine concentrations showed a

Table 6
Cannabinoids Detected in Human Brain

Reference	Number of cases	Brain regions defined	Extraction technique	Detection technique	Deuterated standards
Giroud et al. *(45)*	1	+	SPE	GC–MS	+

–, no; +, yes.

Table 7
Phencyclidine Detected in Human Brain

Reference	Number of cases	Brain regions defined	Extraction technique	Detection technique	Deuterated standards
Budd et al. *(75)*	11	–	n.d.	GC–NPD	n/a

–, no; +, yes; n/a, not applicable; n.d., not defined.

mean of 0.28 µg/g and a range of 0–2.5 µg/g. The study focused on the ratio of unconjugated morphine to total morphine in the blood, and therefore, no conclusions were drawn from the total morphine concentrations in the brain specimens *(28)*. To establish possible patterns or relationships, these data were analyzed in a follow-up study in 1989 using artificial intelligence computer software. All programs applied in this study found total morphine useful in diagnosing morphine intoxication if the concentration in the brain was greater than 0.08 µg/g or greater than the concentration of unconjugated morphine in the blood. Short, moderate, and long-time intervals between last dose and death were differentiated. In addition to other factors, rapid deaths were characterized by total morphine concentrations in the brain greater than 0.16–0.22 µg/g. The author concluded that, for cases with a known history, the artificial intelligence programs were successful 70–90% of the time in classifying the case according to response and time *(77)*.

Pare et al. investigated whether regional morphine concentrations in the brain correlate with opiate receptor density. The following brain sections from 21 suspected heroin-associated fatalities were analyzed for total morphine: brain stem (medulla , pons, and midbrain), cerebellum, cerebral cortex (basal ganglion), hypothalamus, and the thalamus. In all of the cases where death was attributed to narcotic intoxication, the total morphine concentration exceeded 0.2 µg/g in one or more of the brain sections. In more than 50% of the specimens analyzed, the highest total morphine concentration was found in the brain stem or the thalamus – both regions have high concentrations of opiate receptors. The authors concluded that the total morphine concentration in the thalamus could be used to predict concentration in the blood *(27)*.

Based on investigations of 62 cases of intoxication with opiates where information about the survival time was available, Vycudilik *(55)* found that comparison of the unconjugated morphine concentration in the medulla and the cerebellum provided information on the interval between administration and death. Calculating the ratio between the unconjugated morphine concentration in the medulla and the cerebellum resulted in three categories: A ratio less than one indicated short survival time (<1 h between administration of opiates and death); a ratio value of approximately one indicated a moderate survival time (1–2 h between administration of opiates and death); and a ratio greater than one (approximately two or more) indicated a long survival time (more than 6 h between administration of opiates and death). A similar correlation was observed for 6-acetylmorphine. The analysis of the unconjugated morphine concentration in the medulla provided information about consumption of opiates even when morphine was no longer detectable in the blood *(55)*.

5.2. Cocaine

In the human body cocaine is rapidly metabolized by esterases and spontaneously hydrolyzed at physiological pH to form benzoylecgonine, ecgonine methyl ester and, through N-demethylation, norcocaine. Norcocaine possesses pharmacological activity comparable to cocaine. In cases where cocaine is used in combination with ethanol, cocaethylene is also found as an active metabolite, but its role in the fatal outcome of such cases is not totally understood *(78)*. Nevertheless, cocaethylene has been detected in brain, and because of its longer half-life, cocaethylene concentrations can exceed the concentration of cocaine *(41,53)*.

Numerous publications were found reporting concentrations of cocaine (and cocaine metabolites) in brain specimens, describing a wide concentration range and thus making interpretation difficult. The question of distribution of cocaine within the brain, as well as its major metabolite, benzoylecgonine, was addressed in several studies. Spiehler et al. reviewed 37 cocaine intoxication cases and 46 cases in which cocaine was incidental to the cause of death. In two of the cases, cocaine and benzoylecgonine were evenly distributed throughout the brain (in one case, the cocaine was injected and in the other, snorted), with similar concentrations found in dopamine-rich (substantia nigra) and dopamine-poor brain regions (frontal and occipital cortex, cerebellum, medulla, and spinal cord). Cases of fatal intoxication showed a mean cocaine concentration of 13.3 µg/g in the brain with a wide range of 0.17–31.0 µg/g; in cases where cocaine was incidental to the cause of death, the mean cocaine concentration was 0.12 µg/g in the brain with a range of 0–0.7 µg/g.

Cocaine readily crosses the blood–brain barrier; benzoylecgonine, however, is restricted. When significant amounts are detected in the brain, it would appear to be derived from cocaine. Therefore, the ratio between cocaine and benzoylecgonine in brain specimens for cases of fatal intoxication (mean 14.7) was different than that found in incidental cases (mean 0.87). The corresponding ranges overlapped, 0.12–100 for fatal intoxications and 0.04–6 for incidental cases. The ratio between brain and blood concentration for cocaine and benzoylecgonine provided information as to when the drug was used. In fatalities related to intoxication, the mean ratio between brain and blood concentration for cocaine was 9.60 (range: 0.65–155, median: 3.8) compared to 2.5 for incidental cases (range: 0.6–9.2). For benzoylecgonine, the mean ratio between brain and blood concentrations for cases of fatal intoxication was 0.36 (range: 0.04–1.0, median: 0.38) compared to 1.4 for incidental cases (range: 0.05–6.5). Again, there was an overlap in the corresponding ranges, and the high ratio of 1.4 for benzoylecgonine in incidental cases was attributed to accumulation of the metabolite in the brain tissue after chronic cocaine abuse *(38)*.

In another study of 10 cases, the ratio between brain and blood concentrations for cocaine was lower in cases where cocaine intoxication was determined as the cause of death than for cases where death was attributed to excited delirium (ratios >4). These observations differed from the trend noted by Spiehler *(38)* even though all ratios for cases of cocaine intoxication fell within the previously published range *(53)*. Different regions of the brain were examined in two cocaine intoxication cases and in one additional case in which a gunshot wound was the cause of death. In each case, five to eight different brain regions were investigated (basal ganglia, cerebellum, cerebral white, cerebral gray, hypothalamus, hippocampus, motor cortex, frontal cortex, lenticular nuclei, thalamus, occipital cortex, pons, temporal cortex, and medulla). Contrary to previous findings *(38)*, the cocaine and benzoylecgonine distribution throughout the brain was dissimilar. For example, the region of the brain containing the highest concentration of cocaine differed in all three cases that were included in the study, namely the hippocampus, the thalamus, and the basal ganglia *(18)*.

In a larger study, the concentration of cocaine and its major metabolites was determined in 15 different brain regions of 14 chronic cocaine users, including subjects who had recently ingested a high dose of cocaine as well as subjects who did not use the drug immediately before death. Because cocaine has a high affinity to dopamine transporters, dopamine-rich areas of the brain were included. The study also included dopamine-transporter-poor areas of the brain. The following regions of the brain were investigated: frontal cortex, temporal cortex, occipital cortex, parietal cortex, cingulated cortex, cerebellar cortex, caudate, putamen, parolefactory cortex, internal globus pallidus, external globus pallidus, Ammon's horn of hippocampus, hypothalamus, medial-dorsal thalamus, and medial pulvinar thalamus. Although the concentrations of cocaine and its metabolites in the brain varied markedly among the chronic users of the drug, they showed little regional heterogeneity. It could not be clarified whether an initial selective binding of cocaine to the striatal dopamine transporter occurred in vivo followed by rapid redistribution to other brain areas or whether an initial non-selective distribution of cocaine to all areas of the brain was the reason for the relatively homogenous regional distribution pattern. Moreover, some redistribution between death and freezing of the brain specimens could not be excluded *(60)*.

5.3. Amphetamines

Little data exist reporting concentrations of amphetamine *(46,63,70)* and 3,4-MDEA *(64,74)* in brain specimens. Interpretation of results has not been addressed. The regional distribution within the brain has been investigated for methamphetamine as well as 3,4-MDMA and 3,4-methylenedioyxamphetamine (MDA). Concentrations of methamphetamine in brain specimens of autopsy cases vary with a range of 0.02–20.9 µg/g *(49)*.

In a larger study, the concentration of methamphetamine and its metabolite, amphetamine, was determined in 15 brain regions of 14 chronic methamphetamine users. Dopamine-rich areas as well as dopamine-poor areas of the brain were investigated including the parietal cortex, frontal cortex, occipital cortex, temporal cortex, cingulated cortex, white matter, cerebellar cortex, Ammon's horn of hippocampus, putamen, caudate, internal globus pallidus, external globus pallidus, hypothalamus, medial-dorsal thalamus, and medial pulvinar thalamus. Although the concentrations of methamphetamine in the brain varied markedly among chronic users of the drug (0.24–56.6 µg/g), methamphetamine was distributed homogeneously within the brain. The authors concluded that any marked preferential uptake or retention of methamphetamine or amphetamine in dopamine-rich (striatum) versus dopamine-poor brain areas of chronic methamphetamine users seemed unlikely. However, any redistribution occurring after death and before freezing of the brain specimens could not be excluded *(50)*.

To study the regional distribution of MDMA and its metabolite, MDA, in a fatal intoxication case, different areas of the brain (temporal cortex, parietal cortex, frontal cortex, occipital cortex, cerebellum, and brainstem) were investigated. Regional differences were found in the various brain regions with the highest concentrations of MDMA as well as MDA being in the frontal and parietal cortex *(51)*.

5.4. Cannabinoids

One publication reporting the concentration of cannabinoids in the brain was found in the literature. The concentration in the cortex of delta-9-THC was 0.014 µg/g, for 11-hydroxy-delta-9-tetrahydrocannabinol (11-OH-THC) it was 0.012 µg/g, and for 11-nor-delta-9-tetrahydrocannabinol carboxylic acid (THC-COOH) it was 0.024 µg/g. The authors of this case report noted that concentrations in the brain might be less influenced by the release of cannabinoids from fat compartments than concentrations in blood *(45)*.

5.5. Phencyclidine

A single publication describing the concentration of PCP in 11 brain specimens was found in the literature. The concentration of PCP in the brain ranged from 0.03 to 0.81 µg/g. Because PCP has a very large volume of distribution, the correlation coefficient between blood and brain concentration of 0.26 found in this study indicated incomplete distribution of PCP throughout the body in some of the cases *(75)*.

6. Advantages and Disadvantages of Brain as a Drug-Testing Matrix

In forensic toxicology, conclusions about the effects of a drug-of-abuse on the deceased have to be drawn from the results of postmortem specimens. These conclusions are based either on reference data generated by investigators in the testing laboratories or the literature (animal studies, clinical studies, systematic postmortem studies, case reports, etc.). This task requires trained experts with experience because there are numerous practical problems that have to be taken into consideration.

The drug concentrations found in postmortem cases span a wide range, reflecting first-time users as well as subjects with tolerance to the drug. Moreover, inherent physiological differences between individuals must be considered, and specific case details – which could explain the individual circumstances under which death occurred – could be either absent or misleading. In drug abuse cases, multiple drug use is frequently observed, resulting in complex interactions that make the drug effects difficult to predict. Therefore, all measurable drugs should be considered significant when determining the cause of death.

The majority of published extraction procedures and data collections for drugs-of-abuse utilized serum, plasma, or blood specimens. In postmortem forensic toxicology, however, the isolated investigation of a blood specimen is often insufficient in determining whether a drug-of-abuse caused the death of an individual. This may be demonstrated for cocaine-related deaths, where blood concentrations of cocaine are indistinguishable from concentrations in recreational users *(79)*. Blood concentrations of morphine in heroin-related deaths also overlap with those found in non-drug-related deaths *(80)*. Furthermore, blood concentrations of methadone in fatalities are frequently below the plasma concentrations targeted in clinical practice *(81)*. Moreover, the phenomenon of postmortem drug redistribution into the blood from reservoirs with a high concentration, which results in site-dependent differences and time dependent changes, is well documented in the literature for methadone, cocaine, benzoylecgonine, and cocaethylene *(81–83)*.

To provide the appropriate foundation on which an expert opinion can be rendered in fatal cases of drug abuse, additional tissue analyses are often required. A variety of tissue specimens have been used in postmortem forensic toxicology *(19)*.

Brain specimens have some advantages over all other tissues in determining cause of death. In respect to the decomposition of postmortem specimens in forensic cases, brain specimens show greater stability and retarded putrefaction compared to other tissues or blood *(84)*. In addition, following death, bacteria that possess enzymes, which add to the decomposition of labile molecules, transmigrate from the gastrointestinal tract and the lungs throughout the body. Because the brain is a separate compartment, the effects of these bacteria upon the brain are delayed. Also, metabolic activity is lower in the brain than in other tissues or blood. Moriya et al. demonstrated that cocaine was stable in decomposed homogenates of human brain at 20–25 °C as well as at 37 °C over an observed period of 24 h. In specimens of rabbit brains, cocaine degraded much more slowly than in blood or liver over a period of 5 days *(44)*. These factors increase the likelihood of the detection of drugs-of-abuse in brain specimens compared with other tissues, especially when the drugs possess low stability (e.g., cocaine).

Drugs-of-abuse establish their effects, including fatal side effects such as respiratory depression, through the CNS. It can be assumed that concentrations of drugs-of-abuse measured in postmortem brain specimens should be close or equal to perimortem concentrations of the drug at its site of action.

The superiority of brain specimens over other tissues has been demonstrated in numerous publications for opiates and for cocaine and its metabolites. In the case of opiates, the likelihood of detecting 6-acetylmorphine as evidence of heroin consumption was higher in human brain compared with other tissues and blood *(56)*. Moreover, the presence of 6-acetylmorphine in the brain was used as an indicator of a short survival time. A total morphine concentration in the brain that is greater than the concentration of unconjugated morphine in the blood was defined as a possible indicator of acute intoxication. Rapid deaths were characterized by total morphine concentrations in the brain that were greater than 0.16–0.22 µg/g. However, a correlation between morphine distribution in different brain regions and proposed sites of action could not be established though frequently the highest concentrations of total morphine were found in the brainstem and the thalamus – areas with high concentrations of opiate receptors. Comparison of unconjugated morphine concentrations in the medulla and the cerebellum provided information on the interval between administration of opiates and death although this time-dependent concentration gradient between two parts of the brain is currently unexplained.

Similar to opiates, when establishing cocaine intoxication, brain specimens appear to be superior to other tissue specimens or blood. The mean cocaine concentration in the brain after fatal intoxication was clearly different from cases where cocaine was incidental to the cause of death; nevertheless, the observed ranges overlapped *(38)*. The same phenomenon was observed for the ratio between cocaine and benzoylecgonine in brain specimens and, to a lesser degree, for the ratio between brain and blood concentrations of cocaine and benzoylecgonine. These ratios provided information on the interval between administration of cocaine and death, the latter with a proposed peak brain/blood cocaine ratio of approximately 10 between 1 and 2 h after cocaine administration *(38)*. A high concentration of benzoylecgonine in the brain due to accumulation of this metabolite could be used as an indicator of chronic cocaine use.

Throughout the brain of chronic cocaine users, distribution of cocaine and benzoylecgonine appeared to be uniform and did not correlate with proposed sites of action.

Similar results (homogeneous drug distribution) could be found for chronic methamphetamine users; in contrast, MDMA and MDA showed regional differences in brain distribution (highest concentrations in frontal and parietal cortex).

Although reliable procedures for the extraction of complex matrices such as brain specimens are available today and have been standardized through the use of automated devices (especially for SPE), relatively few papers on drugs-of-abuse in brain specimens have been published recently. Modern analytical instruments, in particular the combination of chromatographic techniques with MS, offer sufficient selectivity, sensitivity, and produce precise quantitative results. When applying GC–MS or LC–MS, stable isotopes as internal standards added at the earliest stage of sample preparation increase reproducibility. Therefore, comparable results are produced, which could be used in the future to create a data collection of reliable reference values to enhance accurate interpretation of results. Moreover, these procedures can be successfully applied to small amounts of brain specimen, also allowing for the detection of drugs-of-abuse in the small structures of the brain. In combination with recently developed immunohistochemical techniques to determine the topographic distribution of the receptors as well as the drugs, a range of opportunities for the systematic investigation of drugs-of-abuse in postmortem human brain specimens are now available *(85–88)*.

REFERENCES

1. Yew DT, Kwong WH, Yu MC. *Basic Neuroanatomy*. Singapore, New Jersey, London, Hong Kong: World Scientific Publishing Co. Pte. Ltd., 1996.
2. Cremer HD, Koerper-und Zellbestandteile. In Rauen HM. *Biochemisches Taschenbuch*. Berlin, Goettingen, Heidelberg: Springer Verlag, 1956:831–832 (German).
3. Leuthardt F. Lehrbuch der physiologischen Chemie, 15. *Auflage*. Berlin: Walter de Gruyter & Co, 1963:662–663 (German).
4. Bigeon A, Dillon K, Volkow ND, Hitzemann RJ, Fowler JS, Wolf AP. Quantitative autoradiography of cocaine binding sites in human brain postmortem. *Synapse* 1992;10:126–130.
5. Maurer R, Cortes R, Probst A, Palacios JM. Multiple opiate receptor in human brain: an autoradiographic investigation. *Life Sci* 1983;33:231–234.
6. Musshoff F, Schmidt P, Madea B. Forensische Untersuchungen zur Suchtgenese. In: Madea B, Brinkmann B, ed. *Handbuch gerichtliche Medizin, Band 2*. Berlin, Heidelberg, New York: Springer Verlag, 2003:678 (German).
7. Gabilondo AM, Meana JJ, Barturen F, Sastre M, Garcia-Sevilla JA. μ-Opioid receptor and α_2-adrenoceptor agonist binding sites in the postmortem brain of heroin addicts. *Psychopharmacology (Berl)* 1994;115:135–140.
8. Herkenham M, Lynn AB, Little MD, Johnson MR, Melvin LS, De Costa BR, Rice KC. Cannabinoid receptor localization in brain. *Proc Natl Acad Sci USA* 1990;87:1932–1936.
9. Oldendorf WH, Hyman S, Braun L, Oldendorf Z. Blood-brain barrier: penetration of morphine, codeine, heroin, and methadone after carotid injection. *Science* 1972;178:984–986.
10. Bjoerkman S, Akeson J, Helfer M, Fyge A, Gustafsson LL. Cerebral uptake of morphine in the pig calculated from arterio-venous plasma concentration gradients: an alternative to tissue microdialysis. *Life Sci* 1995;57(25):2335–2345.
11. Mullis KB, Perry DC, Finn AM, Stafford B, Sadee W. Morphine persistence in rat brain and serum after single doses. *J Pharmacol Exp Ther* 1979;208(2):228–231.
12. Wu D, Kang YS, Bickel U, Pardridge WM. Blood-brain barrier permeability to morphine-6-glucuronide is markedly reduced compared with morphine. *Drug Metab Dispos* 1997;25(6):768–771.
13. Som P, Oster ZH, Wang GJ, Volkow ND, Sacker DF. Spatial and temporal distribution of cocaine and effects of pharmacological interventions: whole body autoradiographic microimaging studies. *Life Sci* 1994;55(17):1375–1382.
14. Mule SJ, Casella GA, Misra AL. Intracellular disposition of ^{3}H-cocaine, ^{3}H-norcocaine, ^{3}H-benzoylecgonine and ^{3}H-benzoylnorecgonine in the brain of rats. *Life Sci* 1976;19:1585–1596.
15. Nahas G, Leger C, Tocque B, Hoellinger H. The kinetics of cannabinoid distribution and storage with special reference to the brain and testis. *J Clin Pharmacol* 1981;21:208S–214S.
16. Kim C, Speisky MB, Kalant H. Simultaneous determination of biogenic amines and morphine in discrete rat brain regions by high-performance liquid chromatography with electrochemical detection. *J Chromatogr* 1986;370:303–313.

17. De Zeeuw RA. Drug screening in biological fluids: the need for a systematic approach. *J Chromatogr B* 1997;689:71–79.
18. Browne SP, Moore CM, Scheurer J, Tebbett IR, Logan BK. A rapid method for the determination of cocaine in brain tissue. *J Forensic Sci* 1991;36(6):1662–1665.
19. Drummer OH, Gerostamoulos J. Postmortem drug analysis: analytical and toxicological aspects. *Ther Drug Monit* 2002;24:199–209.
20. Stimpfl T, Vycudilik W. Automatic screening in postmortem toxicology. *Forensic Sci Int* 2004;142:115–125.
21. Sprague GL, Takemori AE. Improved method for morphine extraction from biological samples. *J Pharm Sci* 1979;68(5):660–662.
22. Robinson AE, Williams FM. Post-mortem distribution of morphine in heroin addicts. *Med Sci Law* 1971;11:135–138.
23. Reed D, Spiehler VR, Cravey RH. Two cases of heroin-related suicide. *Forensic Sci* 1977;9(1):49–52.
24. Spiehler VR, Cravey RH, Richards RG, Elliott HW. The distribution of morphine in the brain in fatal cases due to the intravenous administration of heroin. *J Anal Toxicol* 1978;2:62–67.
25. Reed D. Comparison of Spectrofluorometric and GC/MS procedures for the quantitation of morphine in blood and brain. *Clin Toxicol* 1979;14(2):160–180.
26. Raffa RB, O'Neill JJ, Tallarida RJ. Rapid extraction and measurement of morphine and opiate antagonists from rat brain using high-performance liquid chromatography and electrochemical detection. *J Chromatogr A* 1982;238(2):515–519.
27. Pare EM, Monforte JR, Thibert RJ. Morphine concentrations in brain tissue from heroin-associated deaths. *J Anal Toxicol* 1984;8:213–216.
28. Spiehler VR, Brown R. Unconjugated morphine in blood by radioimmunoassay and gas chromatography/mass spectrometry. *J Forensic Sci* 1987;32(4):906–916.
29. Kintz P, Mangin P, Lugnier AA, Chaumont AJ. Toxicological data after heroin overdose. *Hum Toxicol* 1989;8(6):487–489.
30. Borg PJ, Sitaram BR, Taylor DA. Ion-pair extraction and liquid chromatographic analysis of morphine in rat brain and plasma. *J Chromatogr B* 1993;621(2): 165–172.
31. Sticht G, Kaeferstein H, Schmidt P. Zur Wertigkeit des Nachweises von 6-Acetylmorphin in Leichenorganen. *Toxichem+Krimtech* 1993;60(2):34–38 (German).
32. Moriya F, Hashimoto Y. Distribution of free and conjugated morphine in body fluids and tissues in fatal heroin overdose: is conjugated morphine stable in postmortem specimens? *J Forensic Sci* 1997;42(4):736–740.
33. Norheim G. Methadone in autopsy cases. *Z Rechtsmed* 1973;73:219–224.
34. Lundberg GD, Garriott JC, Reynolds PC, Cravey RH, Shaw RF. Cocaine-related death. *J Forensic Sci* 1976;22:402–408.
35. Chinn DM, Crouch DJ, Peat MA, Finkle BS, Jennison TA. Gas chromatography-chemical ionization mass spectrometry of cocaine and its metabolites in biological fluids. *J Anal Toxicol* 1980;4(1):37–42.
36. Griesemer EC, Liu Y, Budd RD, Raftogianis L, Noguchi TT. The determination of cocaine and its major metabolite, benzoylecgonine, in postmortem fluids and tissues by computerized gas chromatography/mass spectrometry. *J Forensic Sci* 1983;28(4):894–900.

37. Poklis A, Mackell MA, Graham M. Disposition of cocaine in fatal poisoning in man. *J Anal Toxicol* 1985;9:227–229.
38. Spiehler VR, Reed D. Brain concentrations of cocaine and benzoylecgonine in fatal cases. *J Forensic Sci* 1985;30(4):1003–1011.
39. Poklis A, Maginn D, Barr JL. Tissue disposition of cocaine in man: a report of five fatal poisonings. *Forensic Sci Int* 1987;33(2):83–88.
40. Mittleman RE, Cofino JC, Hearn WL. Tissue distribution of cocaine in a pregnant woman. *J Forensic Sci* 1989;34(2):481–486.
41. Hime GW, Hearn WL, Rose S, Cofino J. Analysis of cocaine and cocaethylene in blood and tissues by GC-NPD and GC-ion trap mass spectrometry. *J Anal Toxicol* 1991;15(5):241–245.
42. Patrick KS, Boggan WO, Miller SR, Middaugh LD. Gas chromatographic-mass spectrometric determination of plasma and brain cocaine in mice. *J Chromatogr B* 1993;621(1):89–94.
43. Moriya F, Hashimoto Y. The effect of postmortem interval on the concentrations of cocaine and cocaethylene in blood and tissues: an experiment using rats. *J Forensic Sci* 1996;41(1):129–133.
44. Moriya F, Hashimoto Y. Postmortem stability of cocaine and cocaethylene in blood and tissue of humans and rabbits. *J Forensic Sci* 1996;41(4):612–616.
45. Giroud C, Michaud K, Sporkert F, Eap C, Augsburger M, Cardinal P, Mangin P. A fatal overdose of cocaine associated with coingestion of marijuana, buprenorphine, and fluoxetine. Body fluid and tissue distribution of cocaine and its metabolites determined by hydrophilic interaction chromatography-mass spectrometry (HILIC-MS) *J Anal Toxicol* 2004;28(6):464–474.
46. Kojima T, Une I, Yashiki M. CI-mass fragmentographic analysis of methamphetamine and amphetamine in human autopsy tissues after acute methamphetamine poisoning. *Forensic Sci Int* 1983;21:253–258.
47. Kojima T, Une I, Yashiki M, Noda J, Sakai K, Yamamoto K. A fatal methamphetamine poisoning associated with hyperpyrexia. *Forensic Sci Int* 1984;24: 87–93.
48. Terada M. Determination of methamphetamine and its metabolites in rat tissues by gas chromatography with a nitrogen-phosphorus detector. *J Chromatogr* 1985;318(2):307–318.
49. Hara K, Nagata T, Kimura K. Forensic toxicologic analysis of methamphetamine and amphetamine in body materials by gas chromatography/mass spectrometry. *Z Rechtsmed* 1986;96:93–104.
50. Kalasinsky KS, Bosy TZ, Schmunk GA, Reiber G, Anthony RM, Furukawa Y, Guttman M, Kish SJ. Regional distribution of methamphetamine in autopsied brain of chronic human methamphetamine users. *Forensic Sci Int* 2001;116:163–169.
51. De Letter EA, Clauwaert KM, Lambert WE, Van Bocxlaer JF, De Leenheer AP, Piette MHA. Distribution study of 3,4-methylenedioxymethamphetamine and 3,4-methylenedioxyamphetamine in a fatal overdose. *J Anal Toxicol* 2002;26:113–118.
52. Scheurer J, Moore CM. Solid-phase extraction of drugs from biological tissues – a review. *J Anal Toxicol* 1992;16:264–269.
53. Hernandez A, Andollo W, Hearn WL. Analysis of cocaine and metabolites in brain using solid phase extraction and full-scanning gas chromatography/ion trap mass spectrometry. *Forensic Sci Int* 1994;65(3):149–156.

54. Moore C, Browne S, Tebbett I, Negrusz A. Determination of cocaine and its metabolites in brain tissue using high-flow solid-phase extraction columns and high-performance liquid chromatography. *Forensic Sci Int* 1992;53(2):215–219.

55. Vycudilik W. Comparative morphine determination in brain segments by GC/MS. A means of determining the survival time. *Z Rechtsmed* 1988;99(4):263–272.

56. Goldberger BA, Cone EJ, Grant TM, Caplan YH, Levine BS, Smialek JE. Disposition of heroin and its metabolites in heroin-related deaths. *J Anal Toxicol* 1994;18(1):22–28.

57. Klingmann A, Skopp G, Pedal I, Poetsch L, Aderjan R. Distribution of morphine and morphine glucuronides in body tissues and body fluids. Postmortem findings in short survival time. *Arch Kriminol* 2000;206(1–2):38–49.

58. Stimpfl T, Jurenitsch J, Vycudilik W. General unknown screening in postmortem tissue and blood samples: a semi-automatic solid-phase extraction using polystyrene resins followed by liquid–liquid extraction. *J Anal Toxicol* 2001;25(2):125–129.

59. Stichenwirth M, Stellwag-Carion C, Klupp N, Hoenigschnabl S, Vycudilik W, Bauer G, Risser D. Suicide of a body packer. *Forensic Sci Int* 2000;108:61–66.

60. Kalasinsky KS, Bosy TZ, Schmunk GA, Ang L, Adams V, Gore SB, Smialek J, Furukawa Y, Guttman M, Kish SJ. Regional distribution of cocaine in postmortem brain of chronic human cocaine users. *J Forensic Sci* 2000;45(5):1041–1048.

61. Shimomura ET, Hodge GD, Paul BD. Examination of postmortem fluids and tissues for the presence of methylecgonidine, ecgonidine, cocaine, and benzoylecgonine using solid-phase extraction and gas chromatography-mass spectrometry. *Clin Chem* 2001;47(6):1040–1047.

62. Furnari C, Ottaviano V, Sacchetti G, Mancini M. A fatal case of cocaine poisoning in a body packer. *J Forensic Sci* 2002;47(1):208–210.

63. Meyer E, Van Bocxlaer JF, Dirinck IM, Lambert WE, Thienpont L, De Leenheer AP. Tissue distribution of amphetamine isomers in a fatal overdose. *J Anal Toxicol* 1997;21:236–239.

64. Weinmann W, Bohnert M. Lethal monointoxication by overdosage of MDEA. *Forensic Sci Int* 1998;91:91–101.

65. Garcia-Repetto R, Moreno E, Soriano T, Jurado C, Gimenez MP, Menendez M. Tissue concentrations of MDMA and its metabolite MDA in three fatal cases of overdose. *Forensic Sci Int* 2003;135:110–114.

66. Annesley TM. Ion suppression in mass spectrometry. *Clin Chem* 2003;49(7):1041–1044.

67. Richards RG, Reed D, Cravey RH. Death from intravenously administered narcotics: a study of 114 cases. *J Forensic Sci* 1976;21:467–482.

68. Ziminski KR, Wemyss CT, Bidanset JH, Manning TJ, Lukash L. Comparative study of postmortem barbiturates, methadone, and morphine in vitreous humor, blood, and tissue. *J Forensic Sci* 1984;29(3):903–909.

69. Pollak S, Vycudilik W, Mortinger H. Toedliche Heroinvergiftung durch intrakorporalen Drogentransport (Body-Packing). *Med Sach* 1988;84:167–169 (German).

70. Heinemann A, Miyaishi S, Iwersen S, Schmoldt A, Pueschel K. Body-packing as cause of unexpected sudden death. *Forensic Sci Int* 1998;92:1–10.

71. Morild I, Stajic M. Cocaine and fetal death. *Forensic Sci Int* 1990;47:181–189.

72. Rohrig TP, Prouty RW. Tissue distribution of methylenedioxymethamphetamine. *J Anal Toxicol* 1992;16:52–53.

73. Katsumata S, Sato K, Kashiwade H, Yamanami S, Zhou H, Yonemura I, Nakajima H, Hasekura H. Sudden death due presumably to internal use of methamphetamine. *Forensic Sci Int* 1993;62:209–215.

74. Fineschi V, Centini F, Mazzeo E, Turillazzi E. Adam (MDMA) and Eve (MDEA) misuse: an immunohistochemical study on three fatal cases. *Forensic Sci Int* 1999;104:65–74.

75. Budd RD, Liu Y. Phencyclidine concentrations in postmortem body fluids and tissues. *J Toxicol Clin Toxicol* 1982;19(8):843–850.

76. Way EL, Young JM, Kemp JW. Metabolism of heroin and its pharmacologic implications. *Bull Narcotics* 1965;17:25–33.

77. Spiehler VR. Computer-assisted interpretation in forensic toxicology: morphine-involved deaths. *J Forensic Sci* 1989;34(5):1104–1115.

78. Karch SB. Karch's pathology of drug abuse, 3rd ed., Boca Raton, London, New York, Washington DC: CRC Press LLC, 2002:39–43.

79. Jenkins AJ, Levine B, Titus J, Smialek JE. The interpretation of cocaine and benzoylecgonine concentrations in postmortem cases. *Forensic Sci Int* 1999;101(1):17–25.

80. Garriott JC, Sturner WQ. Morphine concentrations and survival periods in acute heroin fatalities. *N Engl J Med* 1973;289:1276–1278.

81. Milroy CM, Forrest AR. Methadone deaths: a toxicological analysis. *J Clin Pathol* 2000;53(4):277–281.

82. Hearn WL, Keran EE, Wei H, Hime G. Site-dependent postmortem changes in blood cocaine concentrations. *J Forensic Sci* 1991;36(3):673–684.

83. Logan BK, Smirnow D, Gullberg RG. Lack of predictable site-dependent differences and time-dependent changes in postmortem concentrations of cocaine, benzoylecgonine, and cocaethylene in humans. *J Anal Toxicol* 1997;21(1):23–31.

84. Sawyer WR, Forney RB. Postmortem disposition of morphine in rats. *Forensic Sci Int* 1988;38(3–4):259–273.

85. Wehner F, Wehner HD, Subke J, Meyermann R, Fritz P. Demonstration of morphine in ganglion cells of the hippocampus from victims of heroin overdose by means of anti-morphine antiserum. *Int J Legal Med* 2000;113:117–120.

86. Wehner F, Wehner HD, Schieffer MC, Subke J. Immunohistochemical detection of methadone in the human brain. *Forensic Sci Int* 2000;112:11–16.

87. Schmidt P, Schmolke C, Musshoff F, Prohaska C, Menzen M, Madea B. Numerical density of μ-opioidreceptor expressing neurons in the frontal cortex of drug related fatalities. *Forensic Sci Int* 2001;115:219–229.

88. Karch SB. Alternate strategies for postmortem drug testing. *J Anal Toxicol* 2001;25:393–395.

Index

A

6-Acetylmorphine (6–AM)
  in amniotic fluid, 10
  in brain, 160, 165, 170
  in meconium, 27
  in nails, 59
  in saliva, 91
  in sweat, 107, 111
  in vitreous humor, 125
Advantages of
  amniotic fluid, 5
  brain, 173
  breast milk, 5
  meconium, 34
  nails, 62
Alcohol
  in hair, 76
  in meconium, 36
Alcohol dehydrogenase, 140
Amniotic Fluid
  analysis of, 9
  anatomy, 6
  cocaine, 9
  collection, 8
  pH, 7–8
  physiology, 6
Amphetamines
  in bone, 133–134
  in brain, 165, 168, 172
  in breast milk, 14
  in hair, 73
  in liver, 146

  in meconium, 30–31
  in nails, 50
  in saliva, 90
  in sweat, 110
Antibiotics
  in nails, 43
Antidepressants
  in breast milk, 14
Apocrine glands, 68, 102
Arsenic
  in nails, 43
Atmospheric pressure ionization
        electrospray mass
        spectrometry, 27

B

Benzodiazepines
  in bone, 133–134
  in breast milk, 10–11, 14
  in hair, 73
Benzoylecgonine
  in amniotic fluid, 9
  in bone, 133
  in brain, 170
  in hair, 70
  in liver, 145, 151
  in meconium, 24
  in nails, 51–55
  in saliva, 91
  in sweat, 111
  in vitreous humor, 120, 121

Printed in the United States of America